Statistical Methods and Scientific Inference

BY THE SAME AUTHOR:

Statistical Methods for Research Workers
(1925 — 14th Ed. 1970)

The Design of Experiments
(1935 — 8th Ed. 1971)

The Theory of Inbreeding
(1949, 1965)

Collected Papers of R. A. Fisher (5 Volumes)
Published by the University of Adelaide, 1971-1973

WITH F. YATES:

Statistical Tables
(1938 — 6th Ed. 1963)

Statistical Methods and Scientific Inference

by

Sir Ronald A. Fisher, Sc.D., F.R.S.

D.Sc. (Adelaide, Ames, Chicago, Harvard, Indian Statistical Institute, Leeds, London)
LL.D. (Calcutta, Glasgow)

Honorary Research Fellow, Division of Mathematical Statistics, C.S.I.R.O., University of Adelaide; Foreign Associate, United States National Academy of Sciences; Foreign Honorary Member, American Academy of Arts and Sciences; Foreign Member, American Philosophical Society; Honorary Member, American Statistical Association; Honorary President International Statistical Institute; Foreign Member, Royal Swedish Academy of Sciences; Member, Royal Danish Academy of Sciences; Member, Pontifical Academy; Member, Imperial German Academy of Natural Science; formerly Fellow of Gonville and Caius College, Cambridge; formerly Galton Professor, University of London; and formerly Balfour Professor of Genetics, University of Cambridge.

Hafner Press
A Division of Macmillan Publishing Co., Inc.
New York
Collier Macmillan Publishers
London

First Published 1956
Second Edition, revised . . . 1959
Third Edition, revised and enlarged . 1973

Hafner Press
A Division of Macmillan Publishing Co., Inc.
866 Third Avenue, New York, N.Y. 10022

Collier-Macmillan Canada Ltd., Toronto, Ontario

Library of Congress Catalog Card Number: 72-97990
ISBN: 0-02-844740-9

Printed in U.S.A. by
NOBLE OFFSET PRINTERS, INC.
NEW YORK, N.Y. 10003

NOTE TO THE THIRD EDITION

The Third Edition, 1973, introduces much new material which the author had entered in his interleaved copy of the book for this purpose, some time before his death on 29 July, 1962. This includes a Section on Isaac Todhunter in Chapter II, an extended discussion of the problem involving observations of two kinds in Chapter V, and numerous smaller additions and clarifications throughout the text.

We use Reason for improving the Sciences; whereas we ought to use the Sciences for improving our Reason.

ANTOINE ARNAULD, 1662
(*The Port-Royal Logic*)

Another use to be made of this Doctrine of Chances is, that it may serve in Conjunction with the other parts of the Mathematicks, as a fit Introduction to the Art of Reasoning.

DE MOIVRE, 1718

If logic investigates the general principles of valid thought, the study of arguments, to which it is rational to attach *some* weight, is as much part of it as the study of those which are demonstrative.

J. M. KEYNES, 1921

CONTENTS

CHAPTER I

FOREWORD

The very large output, characteristic of the present time, of works on various aspects of Statistics, many of them of a much higher standard than were available in the past, not all in the scientific field, but covering the requirements of technological, commercial, educational and administrative purposes, is a recent efflorescence following as a natural and perhaps inevitable consequence on the efforts towards abstract understanding largely set on foot long ago by that versatile and somewhat eccentric man of genius, Francis Galton. Although many of the purposes, to which statistical methods and ideas are in our time successfully applied, are not primarily scientific in aim, that is, are not directed specifically towards an improved understanding of the natural world, yet the fruitfulness and success of the train of studies initiated by Galton were, I submit, due to his own outlook of untrammelled scientific curiosity, and to his confidence that it was in regard to scientific problems that a more penetrating statistical methodology was required. Though all branches of statistical science have profited and been revivified by its influence, it is the course of progress achieved on the scientific front which requires recapitulation, if the nature of the whole movement is to be grasped in spite of its growing complexity and diversity.

Galton's great gift lay in his awareness, which grew during his life, of the vagueness of many of the phrases in which men tried to express themselves in describing natural phenomena. He was before his time in his recognition that such vagueness could be removed, and a certain precision of thought attempted by finding quantitative definitions of concepts fit to take the place of such phrases as "the average man", "variability", "the strength of inheritance", and so forth, through the assembly of objective data, and its systematic examination. That the methods he himself used were often extremely crude, and sometimes seriously faulty, is, indeed, the strongest evidence of the eventual value to the progress of science of his unswerving faith that objectivity and rationality were accessible, even in such elusive fields as psychology, if only a factual basis for these qualities were diligently sought. The systematic improvement of statistical methods and the development of their utility in the study of biological variation and inheritance were the aims to which he deliberately devoted his personal fortune, through the support and endowment of a research laboratory under Professor K. Pearson.

The peculiar mixture of qualities exhibited by Pearson made this choice in some respects regrettable, though in others highly successful. Pearson's energy was unbounded. In the course of his long life he gained the devoted service of a number of able assistants, some of whom he did not treat particularly well. He was prolific in magnificent, or grandiose, schemes capable of realization perhaps by an army of industrious robots responsive to a magic wand. In a sense he undoubtedly appreciated Galton's conception of the greatness of the potential contribution of Statistics in the service of Science, and as a means of rendering strictly scientific a range of studies not traditionally included in the Natural Sciences, but, as perceived through his eyes, this greatness was not easily to be distinguished from the greatness of Pearson himself.

The terrible weakness of his mathematical and scientific work flowed from his incapacity in self-criticism, and his unwillingness to admit the possibility that he had anything to learn from others, even in biology, of which he knew very little. His mathematics, consequently, though always vigorous, were usually clumsy, and often misleading. In controversy, to which he was much addicted, he constantly showed himself to be without a sense of justice. In his dispute with Bateson on the validity of Mendelian inheritance he was the bull to a skilful matador. His immense personal output of writings, his great enterprise in publication, and the excellence of production characteristic of the Royal Society and the Cambridge Press, left an impressive literature. The biological world, for the most part, ignored it, for it was indeed both pretentious and erratic. Yet the intrinsic magnitude of some of the problems brought into discussion, the high prestige that mathematical writing always carries, and a certain imaginative boldness, did suffice to save this material from complete neglect. Little as Pearson cared for the past—for example, for the Gaussian tradition—and much as he would have disliked the future of statistical science, his activities have a real place in the history of a greater movement.

Though Pearson did not appreciate it, quantitative biology, especially in its agricultural applications, was beginning to need accurate tests of significance.

So early as Darwin's experiments on growth rate the need was felt for some sort of a test of whether an apparent effect "might reasonably be due to chance". At the same time it was recognized that the available test based on the conventional "probable error" was not always to be relied on. I have discussed this particular case in *The Design of Experiments* (Chapter III).[1] It was characteristic of the early period, and of Pearson, that such difficulties were habitually blamed on "paucity of data", and not ascribed specifically to the fact that mathematicians had so far offered no solution which the practitioner could use, and indeed had not been sufficiently aware of the difficulty to have discussed the problem. As is well known, it was a research chemist, W. S. Gossett, writing under the designation of "Student",[3] who supplied the test which, important as it was in itself, was of far greater importance in inaugurating the first stage of the process by which statistical methods attained sufficient refinement to be of real assistance in the interpretation of data. As a result of his work, problems of distribution were one after another given exact solutions; by about 1930 all statistical problems which were thought to deserve careful treatment were being discussed in terms of mathematically exact distributions, and the tests of significance based upon them. The logical basis of these scientific applications was the elementary one of excluding, at an assigned level of significance, hypotheses, or views of the causal background, which could only by a more or less implausible coincidence have led to what had been observed. The "Theory of Testing Hypotheses" was a later attempt, by authors who had taken no part in the development of these tests, or in their scientific application, to reinterpret them in terms of an imagined process of acceptance sampling, such as was beginning to be used in commerce; although such processes have a logical basis very different from those of a scientist engaged in gaining from his observations an improved understanding of reality.

The exact solutions of the series of problems of distribution left unsolved by the Pearsonian school had not only as its immediate fruit the refinement, by accurate tests of significance, of the experimenter's facility in examining his data critically; at a deeper logical level they allowed of the development of objective principles of estimation, and so revealed the misleading character of many of the methods of estimation commonly advocated. The variety of concepts relevant to the logical basis of a process of estimation of hypothetical quantities by the aid of observational material is considerable, and some account is given of them in Chapter VI. It was early necessary to distinguish Mathematical Likelihood from Mathematical Probability, and the concept of quantity of information (different from the meaning later given to the same phrase in Communication Theory) itself intimately related to the Likelihood, was found to measure effectively the competence of any proposed method of estimation, even in its application to small samples.

From a larger viewpoint than that of merely refining and perfecting the statistical processes used in the examination of a fixed body of data, the concepts of the theory of estimation lent themselves to the effectual comparison of different bodies of data, and therefore of the experimental procedures, or observational programmes capable of giving rise to such observational foundations. This is the leading consideration in the branch of statistical science known as Experimental Design, on which during the last few years very comprehensive and substantial works have appeared. The practical and theoretical study of Experimental Design, with which should be included that of sampling for the purpose of factual ascertainment as developed by Mahalanobis,[2] and by Yates,[4] may be regarded as the second great movement in the development of Statistics for the clarification of scientific thought.

In the author's view, however, there has already appeared a need for the exposition and consolidation of the specifically logical concepts which have emerged as it were as by-products of both (*a*) the purely mathematical elucidation of statistical problems in the first phase, and (*b*) in the second phase the development of experimental designs, logically coherent with the processes used in their discussion, and with the scientific inferences of which they are to supply the basis, so as to form with them a complete illustration of the mode in which new scientific knowledge is generated. Once recognized and applied, there is little danger of such an advance in procedure being lost. Practitioners, however, are not the natural repositories of logical niceties, and teachers especially, engaged in introducing students to these new fields, may value an attempt to consolidate the specifically logical gains of the past half-century, and will perhaps tolerate a certain amount of necessary hair-splitting.

In the introduction to the *Design of Experiments* (p. 9)[1] I have stressed my conviction that the art of framing cogent experiments, like that of their statistical analysis, can each only establish its full significance as parts of a single process of the improvement of natural knowledge; and that the logical coherence of this whole is the only full justification for free individual thought in the formation of opinions applicable to the real world. I would wish again now to reiterate this point of view. To one brought up in the free intellectual atmosphere of an earlier time there is something rather horrifying in the ideological movement represented by the doctrine that reasoning, properly speaking, cannot be applied to empirical data to lead to inferences valid in the real world. It is undeniable that the intellectual freedom that we in the West have taken for granted is now successfully denied over a great part of the earth's surface. The validity of the logical steps by which we can still dare to draw our own conclusions cannot therefore, in these days, be too clearly expounded, or too strongly affirmed.

REFERENCES

1. R. A. Fisher (1935-1966). *The Design of Experiments.* Oliver and Boyd, Edinburgh.
2. P. C. Mahalanobis (1946). On large-scale sample surveys. *Phil. Trans. Roy. Soc.*, B, vol. 231, pp. 329-451.
3. "Student" (1908). The probable error of a mean. *Biometrika*, vol. 6, pp. 1-25.
4. F. Yates (1949). *Sampling Methods for Censuses and Surveys.* Charles Griffin, London.

CHAPTER II

THE EARLY ATTEMPTS AND THEIR DIFFICULTIES

1. Thomas Bayes

For the first serious attempt known to us to give a rational account of the process of scientific inference as a means of understanding the real world, in the sense in which this term is understood by experimental investigators, we must look back over two hundred years to an English clergyman, the Reverend Thomas Bayes, whose life spanned the first half of the eighteenth century. It is indeed only in the present century, with the rapid expansion of those studies which are collectively known as Statistics, that the importance of Bayes' contribution has come to be appreciated. *The Dictionary of National Biography*,[13] representing opinion current in the last quarter of the nineteenth century, does not include his name. The omission is the more striking since this work of reference does include a notice of his father, Joshua Bayes (1671-1746). While the father was no doubt a learned and eloquent preacher, still, in his own time, his son Thomas was for twenty years a Fellow of the Royal Society, and therefore known also as a not inconsiderable mathematician. Indeed, his mathematical contributions to the *Philosophical Transactions* show him to have been in the first rank of independent thinkers, very well qualified to attempt the really revolutionary task opened out by his posthumous paper "An Essay towards solving a problem in the doctrine of chances", which appeared in the *Philosophical Transactions*[1] in 1763, not long after his death in 1761.

It is entirely appropriate that this first attempt should have been made at this time. For more than a century the learned world had been coming to regard deliberate experimentation as the fundamental means to "The Improvement of Natural Knowledge", in the words chosen by the Royal Society. With Isaac Newton, moreover, and such men as Robert Boyle, the possibility of formulating natural law in quantitative terms had been brilliantly exhibited. The nature of the reasoning process by which appropriate inferences, or conclusions, could be drawn from quantitative observational data was ripe for consideration. The prime difficulty lay in the uncertainty of such inferences, and it was a fortunate coincidence that the recognition of the concept of probability, and its associated mathematical laws, in its application to games of chance, should at the same time have provided a possible means by which such uncertainty could be specified and made explicit. In England such a publication as Abraham de Moivre's *Doctrine of Chances*[9] must have been a very immediate stimulus to Bayes' reflexions on this subject.

Bayes' Essay was communicated to the Royal Society some time after his death by his friend Richard Price. Price added various demonstrations and illustrations of the method, and seems to have replaced Bayes' introduction by a prefatory explanation of his own. It is to be regretted that we have not Bayes' own introduction, for it seems clear that Bayes had recognized that the postulate proposed in his argument (though not used in his central theorem) would be thought disputable by a critical reader, and there can be little doubt that this was the reason why his treatise was not offered for publication in his own lifetime. Price evidently laid less weight on these doubts than did Bayes himself; on the other hand he very fully appreciated the importance of what Bayes had done, or attempted, for the advancement of experimental philosophy; although the central theorem of the essay is framed in somewhat academic and abstract terms, without expatiating on the large consequences for human reasoning which would flow from his axiom.

It contains also an element of artificiality, which has obscured its understanding, and which, I believe, is capable of being removed (see Chapter V, Section 6, and reference 6 to this Chapter).

The most important passage of Price's introductory letter is as follows ([1] p. 370):

> In an introduction which he has writ to this Essay, he says, that his design at first in thinking on the subject of it was, to find out a method by which we might judge concerning the probability that an event has to happen, in given circumstances, upon supposition that we know nothing concerning it but that, under the same circumstances, it has happened a certain number of times, and failed a certain other number of times. He adds, that he soon perceived that it would not be very difficult to do this, provided some rule could be found according to which we ought to estimate the chance that the probability for the happening of an event perfectly unknown, should lie between any two named degrees of probability, antecedently to any experiments made about it; and that it appeared to him that the rule must be to suppose the chance the same that it should lie between any two equidistant degrees; which, if it were allowed, all the rest might be easily calculated in the common

method of proceeding in the doctrine of chances. Accordingly, I find among his papers a very ingenious solution of this problem in this way. But he afterwards considered, that the *postulate* on which he had argued might not perhaps be looked upon by all as reasonable; and therefore he chose to lay down in another form the proposition in which he thought the solution of the problem is contained, and in a *scholium* to subjoin the reasons why he thought so, rather than to take into his mathematical reasoning anything that might admit dispute.

The actual mathematics of Bayes' theorem may be expressed very briefly in modern notation.

If in $a+b$ independent trials it has been observed that there have been a successes and b failures, then if p were the hypothetical probability of success in each of these trials, the probability of the happening of what has been observed would be

$$\frac{(a+b)!}{a!\,b!}\,p^a(1-p)^b\,; \tag{1}$$

but if in addition we know, or can properly postulate, that p itself has been chosen by an antecedent random process, such that the probability of p lying in any infinitesimal range dp between the limiting values 0 and 1 is equal simply to

$$dp\,, \tag{2}$$

then the probability of the compound event of p lying in the assigned range, and of the observed numbers of successes and failures occurring will be the product of these two expressions, namely

$$\frac{(a+b)!}{a!\,b!}\,p^a(1-p)^b\,dp\,. \tag{3}$$

But, from the data, such a compound event has happened for some element or other of those into which the total range from 0 to 1 may be divided, so that the probability that any particular one should have happened in fact is the ratio,

$$\frac{\frac{(a+b)!}{a!\,b!}\,p^a(1-p)^b\,dp}{\frac{(a+b)!}{a!\,b!}\int_0^1 p^a(1-p)^b\,dp}, \tag{4}$$

of which the denominator, involving a complete Eulerian integral, is equal $1/(a+b+1)$.

The finite probability that p should lie between any assigned limits u and v may therefore be expressed as the incomplete integral

$$\frac{(a+b+1)!}{a!\,b!}\int_u^v p^a(1-p)^b\,dp\,. \tag{5}$$

The postulate which Bayes regarded as questionable is represented above by the expression (2). The greater part of Bayes' analysis is concerned with approximate forms for the discussion of these integrals, and is of historical rather than mathematical interest for the modern reader. How explicit Bayes is in introducing the critical datum is shown by his own introductory remarks.

The crucial theorem, proposition 8, comes in the second section of the essay, and is preceded by a special explanatory foreword[1] (p. 385):

SECTION II

Postulate 1. I suppose the square table or plane *ABCD* to be so made and levelled, that if either of the balls *O* or *W* be thrown upon it, there shall be the same probability that it rests upon any one equal part of the plane as another, and that it must necessarily rest somewhere upon it.

2. I suppose that the ball *W* shall be first thrown, and through the point where it rests a line *os* shall be drawn parallel to *AD*, and meeting *CD* and *AB* in *s* and *o*; and that afterwards the ball *O* shall be thrown $p+q$ or n times, and that its resting between *AD* and *os* after a single throw be called the happening of the event *M* in a single trial. These things supposed,

Lemma 1. The probability that the point *o* will fall between any two points in the line *AB* is the ratio of the distance between the two points to the whole line *AB*. (The proof occupies two pages, with an examination of incommensurability after the manner of the fifth book of Euclid's elements.)

Lemma 2. The ball *W* having been thrown, and the line *os* drawn, the probability of the event *M* in a single trial is the ratio *Ao* to *AB*.

After a short proof there follow the enunciation of proposition 8 and its demonstration. A single figure is used to represent the square table and the construction upon it, and also outside the square, a graph representing the function,

$$\frac{(a+b)!}{a!\,b!}p^a q^b, \qquad p+q=1,$$

for all values of p from 0 to 1. The latter is used to give geometrical significance to the analytic integrals used above.

In broaching the boundaries of an entirely new field of thought by means of a single illustrative theorem, pregnant as it was, Bayes left untouched many distinctions of importance to its discussion in the future. In respect to the nature of the concept of probability very diverse opinions have been expressed. In particular, although perhaps all would agree that the word denotes a measure of the strength of an opinion or state of judgement, some have insisted that it should properly be used only for the expression of a state of rational judgement based on sufficient objective evidence, while others have thought that equality of probability may be asserted merely from the indifference of, or the absence of *differentiae* in, the objective evidence, if any, and therefore from the total absence of objective evidence, if there were none.

Bayes evidently held the first of these opinions and frames a definition suited—in my view—to show (1) that he was not thinking merely of games of chance, and, (2) at the same time that his concept of probability was that of the mathematicians, such as Montmort[10] and de Moivre, who had treated largely of gambling problems, in which the equality of probability assigned to numerous possible events and combinations of events is a consequence of the assumed perfection of the apparatus and operations employed.

Bayes' definition is[1] (p. 376): "5. The *probability of any event* is the ratio between the value at which an expectation depending on the happening of the event ought to be computed, and the value of the thing expected upon its happening. 6. By *chance* I mean the same as probability." There is no room for doubt that Bayes would have regarded the "expectation" to which he referred as capable of verification to any required approximation, by repeated trials with sufficiently perfect apparatus. Subject to the latent stipulation of fair use, or of homogeneity in the series of tests, his definition is therefore equivalent to the limiting value of the relative frequency of success.

On the contrary Laplace, who needed a definition

wide enough to be used in the vastly diverse applications of the *Théorie analytique*, manifestly inclined to the second view[7] (1820).

La théorie des hasards consiste à réduire tous les évènemens du même genre, à un certain nombre de cas également possibles, c'est-à-dire, tels que nous soyons également indécis sur leur existence; et à déterminer le nombre de cas favorables à l'évènement dont on cherche la probabilité. Le rapport de ce nombre à celui de tous les cas possibles, est la mesure de cette probabilité qui n'est ainsi qu'une fraction dont le numérateur est le nombre des cas favorables, et dont le dénominateur est le nombre de tous les cas possibles.

This differs a little from the form used in 1812:

La théorie des probabilités consiste à réduire tous les évènemens qui peuvent avoir lieu dans une circonstance donnée, à un certain nombre de cas également possibles, c'est-à-dire tels que nous soyons également indécis sur leur existence, et à déterminer parmi ces cas, le nombre de ceux qui sont favorables à l'évènement dont on cherche la probabilité. Le rapport de ce nombre à celui de tous les cas possibles, est la mesure de cette probabilité qui n'est donc qu'une fraction dont le numérateur est le nombre des cas favorables, et dont le dénominateur est celui de tous les cas possibles.

It is seen that Laplace effectively avoids any objective definition, first by using the term *possible* in a context in which *probable* could be used, without explaining what difference, if any, he intends between the two words, and secondly by his indication that equal possibility could be judged without cogent evidence of equality.

In consequence of this difference of concept, Bayes' attempt is exposed to different types of criticism in his own hands and in those of Laplace. While I have, for myself, no doubt that Bayes' definition is the more satisfactory, being not only in accordance with the ideas upon which the *Doctrine of Chances* of his own time was built, but in connecting the comparatively modern notion of *probability*, which seems to have been unknown to the Islamic and to the Greek mathematicians, with the much more ancient notion of an *expectation*, capable of being bought, sold and evaluated, nevertheless it would merely confuse the discussion to give further reasons for this opinion; the difficulties to which Bayes' approach was eventually found to lead can be easily expressed in terms of the notions which he himself favoured.

Whereas Laplace defined probability by means of the enumeration of discrete units, Bayes defined a continuous probability distribution, by a formula for the probability between any pair of assigned limits. He did not, however, consider the metric of his continuum. For in stating his prime postulate in the form that the chance *a priori* of the unknown probability lying between p_1 and p_2 should be equal to p_2-p_1 he might, so far as cogent evidence is concerned, equally have taken any monotonic function of p, such as

$$\phi = \tfrac{1}{2}\cos^{-1}(1-2p), \quad p = \sin^2\phi\,,$$

and postulated that the chance that ϕ should lie between ϕ_1 and ϕ_2 should be

$$\frac{2}{\pi}(\phi_2-\phi_1)\,,$$

so that, instead of inserting the probability *a priori* as

$$dp\,,$$

it would have appeared in the analysis as

$$\frac{2}{\pi}d\phi = \frac{1}{\pi\sqrt{pq}}\,.\,dp\,, \quad q = 1-p\,, \tag{6}$$

a postulate or assumption rather more favourable to extreme values of the unknown p, near to 0 or 1, at the expense of more central values.

Bayes' introduction of an expression representing probability *a priori* thus contained an arbitrary element, and it was doubtless some consciousness of this that led to his hesitation in putting his work forward. He removes the ambiguity formally by means of the auxiliary experiment with the ball W, and in Chapter V it is shown that such an auxiliary experiment can sometimes be realized in practice.

A more important question, however, is whether in scientific research, and especially in the interpretation of experiments, there is cogent reason for inserting a corresponding expression representing probabilities *a priori*. This practical question cannot be answered peremptorily, or in general, for certainly cases can be found, or constructed, in which valid probabilities *a priori* exist, and can be deduced from the data. More frequently, however, and especially when the probabilities of contrasted scientific theories are in question, a candid examination of the data at the disposal of the scientist shows that nothing of the kind can be claimed.

2. George Boole

The superb pre-eminence of Laplace as a mathematical analyst undoubtedly inclined mathematicians for nearly fifty years to the view that the logical approach adopted by him had removed all doubts as to the applicability in practice of Bayes' theorem. That this was indeed Laplace's view may be judged from his reference to the position of Bayes in the history of the subject[7] (p. cxxxvii).

Bayes dans les *Transactions philosophiques* de l'année 1763, a cherché directement la probabilité que les possibilités indiquées par des expériences déjà faites, sont comprises dans des limites données; et il y a parvenu d'une manière fine et très ingénieuse, quoiqu'un peu embarrassée.

I imagine that the hint of criticism in the last phrase is directed against Bayes' hesitation to regard the postulate he required as axiomatic. It will be noticed in the sequel that discussion turned on just this question of its axiomatic nature, and not on the question, more natural to an experimental investigator, of whether, in the particular circumstances of the investigation, the knowledge implied by the postulate was or was not in fact available. It is the submission of the author that actual familiarity with the processes of scientific research helps greatly in the understanding of scientific data, and has in the present century clarified the issue by bringing into prominence the factual question, rather than the abstract question of axiomatic validity. It was not, however, until the weight of opinion among philosophical mathematicians had turned against the supposed axiom that the controversy could come to be examined in this more realistic manner.

Simple examples are provided by genetic situations. In Mendelian theory there are black mice of two genetic kinds. Some, known as homozygotes (*BB*), when mated with brown yield exclusively black offspring; others, known as heterozygotes (*Bb*), while themselves also black, are expected to yield half black and half brown. The expectation from a mating between two heterozygotes is 1 homozygous black, to 2 heterozygotes, to 1 brown. A black mouse from such a mating has thus, prior to any test-mating in which it may be used, a known probability of 1/3 of being homozygous, and of 2/3 of being heterozygous. If, therefore, on testing with a brown mate it yields seven offspring, all being black, we have a situation perfectly analogous to that set out by Bayes in his proposition, and can develop the counterpart of his argument, as follows:

The prior chance of the mouse being homozygous is 1/3; if it is homozygous the probability that the young shall be all black is unity; hence the probability of the compound event of a homozygote producing the test litter is the product of the two numbers, or 1/3.

Similarly, the prior chance of it being heterozygous is 2/3; if heterozygous the probability that the young shall be all black is $1/2^7$, or 1/128; hence the probability of the compound event is the product, 1/192.

But, one of these compound events has occurred; hence the probability after testing that the mouse tested is homozygous is

$$1/3 \div \left(\frac{1}{3} + \frac{1}{192}\right) = 64/65\,,$$

and the probability that it is heterozygous is

$$1/192 \div \left(\frac{1}{3} + \frac{1}{192}\right) = 1/65\,.$$

If, therefore, the experimenter knows that the animal under test is the offspring of two heterozygotes, as would be the case if both parents were known to be black, and a parent of each were known to be brown, or, more strictly, if both had either a brown parent or a brown offspring, cogent knowledge *a priori* would have been available, and the method of Bayes could properly be applied. But, if knowledge of the origin of the mouse tested were lacking, no experimenter would feel he had warrant for arguing as if he knew that of which in fact he was ignorant, and for lack of adequate data Bayes' method of reasoning would be inapplicable to his problem.

It is evidently easier for the practitioner of natural science to recognize the difference between knowing and not knowing than this seems to be for the more abstract mathematician. The traditional line of thought running from Laplace to, for example, Sir Harold Jeffreys in our own time would be to argue that, in the absence of relevant genealogical evidence, there being only two possibilities, mutually exclusive, and with no prior information favouring one rather than the other, it is axiomatic that their probabilities *a priori* are equal, and that Bayes' argument should be applied on this basis. This is to treat the problem, in which we have no genealogical evidence, exactly *as if* the mouse to be tested were known to have been derived from a mating producing half homozygotes and half heterozygotes.

In spite of the high prestige of all that flowed from Laplace's pen, and the great ability and industry of his expositors, it is yet surprising that the doubts which such a process of reasoning from ignorance must engender should begin to find explicit expression only in the second half of the nineteenth century, and then with caution. That extraordinary work,

The Laws of Thought, by George Boole appeared in 1854.[2] Its twentieth chapter is given to problems of causes, and of this the second half to problems in which (p. 320) we "may be required to determine the probability of a particular cause, or of some particular connection among a system of causes, from observed effects,". A hint of Boole's point of view appears in the opening words of Section 20:

> It is remarkable that the solutions of the previous problems are void of any arbitrary element. We should scarcely, from the appearance of the data, have anticipated such a circumstance. It is, however, to be observed, that in all those problems the probabilities of the *causes* involved are supposed to be known *a priori*. In the absence of this assumed element of knowledge, it seems probable that arbitrary constants would *necessarily* appear in the final solution.

The cases chosen by Boole to illustrate this view are two: (*a*) The Reverend J. Michell[8] had calculated that if stars of each magnitude were dispersed at random over the celestial sphere, there would very rarely occur so many apparent double stars or clusters as those actually observed by astronomers. (*b*) The planes of revolution of the planets of the solar system are more nearly coincident than could often occur if these planes had been assigned at random. Instead of treating these calculations, as would now generally be done, as *tests of significance* overthrowing the theory of random dispersal, and therefore all cosmological theories implying random dispersal (disposing of this hypothesis without reference to, or consideration of, any alternative hypothesis which might be actually or conceivably brought forward); instead of this, it had been thought proper to discuss each question as one in inverse probability, and

Boole has no difficulty in showing that as such it requires two elements really unknown, namely the probability *a priori* of random dispersal, and, secondly, the probability, in the aggregate of alternative hypotheses, of the observed frequency of conjunctions being realized. As he says on p. 367, "Any solutions which profess to accomplish this object, either are erroneous in principle, or involve a tacit assumption respecting the above arbitrary elements."

Again in Section 22:

Are we, however, justified in assigning to [these two unknowns] particular values? I am strongly disposed to think that we are not. The question is of less importance in the special instance than in its ulterior bearings. In the received applications of the theory of probabilities, arbitrary constants do not explicitly appear; but in the above, and in many other instances sanctioned by the highest authorities, some virtual determination of them has been attempted. And this circumstance has given to the results of the theory, especially in reference to questions of causation, a character of definite precision, which, while on the one hand it has seemed to exalt the dominion and extend the province of numbers, even beyond the measure of their ancient claim to rule the world; on the other hand has called forth vigorous protests against their intrusion into realms in which conjecture is the only basis of inference. The very fact of the appearance of arbitrary constants in the solutions of problems like the above, treated by the method of this work, seems to imply, that definite solution is impossible, and to mark the point where inquiry ought to stop.

On page 370:

It has been said, that the principle involved in the above and in similar applications is that of the equal distribution of our knowledge, or rather of our ignorance—the assigning to different states of things of which we know nothing, and upon the very ground that we know nothing, equal degrees of probability. I apprehend, however, that this is an arbitrary method of procedure.

And finally, on page 375:

These results only illustrate the fact, that when the defect of data is supplied by hypothesis, the solutions will, in general, vary with the nature of the hypotheses assumed; so that the question still remains, only more definite in form, whether the principles of the theory of probabilities serve to guide us in the election of such hypotheses. I have already expressed my conviction that they do not—a conviction strengthened by other reasons than those above stated . . . Still it is with diffidence that I express my dissent on these points from mathematicians generally, and more especially from one who, of English writers, has most fully entered into the spirit and the methods of Laplace; and I venture to hope, that a question, second to none other in the Theory of Probabilities in importance, will receive the careful attention which it deserves.

These quotations, which I have picked out from the rather lengthy mathematical examples which Boole developed, are sufficient to exhibit unmistakably his logical point of view. He does not, indeed, go so far as to say that no statements in terms of mathematical probability can properly be based on data of the kind considered, but he is entirely clear in rejecting the application to these cases of the method of arriving at such statements which, in the absence of appropriate data, introduced values of the probabilities *a priori* supported only by a questionable axiom.

His phrase, moreover, on supplying by hypothesis what is lacking in the data, points to an abuse very congenial to certain twentieth-century writers.

3. John Venn and the Rule of Succession

An immediate inference from Bayes' theorem assigning the frequency distribution

$$\frac{(a+b+1)!}{a!\,b!}\,p^a\,(1-p)^b\,dp \tag{7}$$

to the probability of success of an event, supposed constant, after a successes have been observed in $a+b$ independent trials, is to calculate the probability of success of a new trial, of the same kind as the others and, like them, independent. For this we have only to multiply the frequency element above by p and integrate between the limits 0 and 1. The result takes the simple form

$$(a+1)/(a+b+2), \tag{8}$$

and this inference came to be known as the Rule of Succession; it is often quoted in the form taken when $b=0$, as leading to the probability $(a+1)/(a+2)$.

It should be emphasized, as it has sometimes passed unnoticed, that such a rule can be based on Bayes' theorem only on certain conditions. It requires that (i) the record of a successes out of $a+b$ trials constitutes the whole of the information available; (ii) the successive trials are independent in the sense that the success or failure of one trial has no effect in favouring the success or failure of subsequent tests, which have in each case the same probabilities.

During the long period over which its correctness was unquestioned, the Rule of Succession had been eagerly seized upon by logicians as providing a solid mathematical basis for inductive reasoning. In his *Logic of Chance*,[14] Venn, who was developing the concept of probability as an objective fact, verifiable by observations of frequency, devotes a chapter to demolishing the Rule of Succession, and this, from a writer of his weight and dignity, had an undoubted effect in shaking the confidence of mathematicians in its mathematical foundation.

Venn, however, does not discuss its foundation, and perhaps was not aware that it had a mathematical basis demonstrated by Laplace; that like other mathematical theorems it contained stipulations specific for its validity; and that in particular it rested upon the supposed, though disputable, axiom used for the demonstration of Bayes' proposition. As in other cases in which a work of demolition is undertaken with great confidence, there is no doubt that Venn in this chapter uses arguments of a quality which he would scarcely have employed had he regarded the matter as one open to rational debate.

After giving instances of not very unreasonable inferences drawn by Laplace and De Morgan with the aid of the Rule, Venn writes[15] (p. 180):

> Let us add an example or two more of our own. I have observed it rain three days successively,—I have found on three separate occasions that to give my fowls strychnine has caused their death,—I have given a false alarm of fire on three different occasions and found the people come to help me each time.

These examples seem to be little more than rhetorical sallies intended to overwhelm an opponent with ridicule. They scarcely attempt to conform with the conditions of Bayes' theorem, or of the rule of succession based upon it. In the last case the reader is to presume, the same neighbours having been deceived on three occasions, that on the fourth they will be for this reason less ready to exert themselves; that is to say, the successive trials are not even conceived to be independent. Objection could be made on the same ground to the first example, which is perhaps particularly unrealistic in that three rainy days are postulated to comprise the whole of the subject's experience of days wet or fine. Perhaps the example could be repaired by making him arrive by air in a region of unknown climate; if so, Bayes' postulate implies that the region has been chosen at random from an aggregate of regions in each of which the probability of rain is constant and independent from day to day, while this probability varies from region to region in an equal distribution from 0 to 1. If applied to cases in which this information is lacking the inference is not indeed ridiculous, though I should agree with Venn that it would often be found to be mistaken, if put to the test of repeated trials, and it can scarcely be doubted that Bayes would have taken the same view. A climate without the unnatural feature of independence of weather from day to day, and therefore without conforming to the conditions of Bayes' theorem, might yet justify the Rule of Succession, in the limited form here used, if the proportion of all rainy days falling in spells of n successive rainy days were

$$\frac{4n}{(n+1)(n+2)(n+3)}\,; \tag{9}$$

and it is clearly a question of ascertainable fact, and

not of personal predilection, whether the climate of any part of the world conforms to such a rule.*

The standardization of drugs by an experimental assay of their potency often involves the determination of the 50% lethal dose, or, that which, with the population of animals sampled for testing, will have a probability of 50% of killing in each case. The rhetorical force of Venn's example lies in the presumption that much more than the 50% lethal dose of strychnine was employed, but a valid criticism of Bayes' theorem through the failure of the Rule of Succession requires a less cavalier treatment of the example. If, for example, 50 strengths of dose were made up at concentrations capable of killing 1%, 3%, . . . , 99% of the animals to be tested, and if each experiment consisted in choosing one of these doses, with equal probability, and applying this dose to each of four hens chosen at random; of ascertaining

* While this simple distribution suffices to justify the Rule of Succession when applied to experience of only wet or of only fine days, the general form of the rule requires that the spells of wet and fine weather must be arranged to fulfil further conditions. Thus the frequency with which two successive spells are of u and v days respectively must be

$$\frac{6(u+1)!\,(v+1)!}{(u+v+3)!}, \tag{10}$$

consistent with the marginal frequency for u and v

$$\frac{12(u!)}{(u+3)!}. \tag{11}$$

Three successive spells of lengths u, v and w, in that order, have frequency

$$\frac{6(v+2)!\,(u+w)!}{(u+v+w+3)!}, \tag{12}$$

while four spells of lengths t, u, v, w will have a frequency

$$\frac{6(t+v+1)!\,(u+w+1)!}{(t+u+v+w+3)!}. \tag{13}$$

first if each of three of these hens had died, and if so of predicting that the fourth would die also, the proportion of successes would agree closely with the fraction 4/5 given by the rule of succession. If, on the contrary, the doses were spaced in toxic content between the 1% and the 99% dosages, either in arithmetical or in geometrical progression, the forecast, though by no means a ridiculous estimate, would doubtless be somewhat in error. *Knowledge* of the experimental conditions might thus justify the rule, though it cannot rationally be based on ignorance of them.

It seems that in this chapter Venn was to such an extent carried away by his confidence that the rule of induction he was criticizing was indefensible in many of its seeming applications, and by his eagerness to dispose of it finally, that he became uncritical of the quality of the arguments he used. A most serious lapse of a general character appears on page 181 at the beginning of Section 9:

> It is surely a mere evasion of the difficulty to assert, as is sometimes done, that the rule is to be employed in those cases only in which we do not know anything beforehand about the mode and frequency of occurrence of the events. The truth or falsity of the rule cannot be in any way dependent upon the ignorance of the man who uses it. His ignorance affects himself only, and corresponds to no distinction in the things.

Taken in its sweeping generality such an argument seems to imply that the extent of the observational data available can have no bearing on the nature or precision of our inferences from them; that a jury ignorant of certain facts ought to give the same verdict as one to whom they have been presented! The precise specification of our knowledge is, however, the same as the precise specification of our ignorance. Certainly, the observer's knowledge or ignorance may have no effect on external objects, but the extent of the observations to which his reasoning is applied does make a selection of those material systems to which he imagines his conclusions to be applicable; and the objective frequencies observed in such selected systems may depend, and indeed must depend if inductive reasoning have any validity, on the observational basis by which the selection is effected, and on which the reasoning is based.

It is certain that Venn understood this in respect of the inductive process generally, and that nothing but inadvertence can have led him to develop, in criticizing "the rule", a mode of argument fatal equally to all inferences based on experience.

Perhaps the most important result of Venn's criticism was the departure made by Professor G. Chrystal in eliminating from his celebrated textbook of *Algebra*[3] the whole of the traditional material usually presented under the headings of *Inverse Probability* and of the *Theory of Evidence.* Chrystal does not discuss the objections to this material, but expresses the opinion that "many of the criticisms of Mr. Venn on this part of the doctrine of chances are unanswerable. The mildest judgement we could pronounce would be the following words of De Morgan himself, who seems, after all, to have 'doubted': 'My own impression derived from this and many other circumstances connected with the analysis of probabilities, is, that the mathematical results have outrun their interpretation.'" (Chapter xxxvi, p. 604.)

It should be noted that De Morgan's remark has

been quoted clean out of its context; that he was not writing on inverse probability, nor even on the theory of evidence, but about the curiously ubiquitous success of methods based on the Normal law of errors, even when applied to cases in which such a law is not accurately plausible. In fact the passage (from the Fourth Appendix, the title of which is "On the average result of a number of observations") goes on:

> and that some simple explanation of the force and meaning of the celebrated integral, whose values are tabulated at the end of this work, will one day be found to connect the higher and lower parts of the subject with a degree of simplicity which will at once render useless (except to the historian) all the works hitherto written.

In reality, the introduction of the inverse method was to De Morgan[11] (p. vi) one of the most important advances to be recorded in the history of the theory of probability. I have already quoted his opinion to this effect, in the introduction to my book on *The Design of Experiments*[5]:

> There was also another circumstance which stood in the way of the first investigators, namely, the not having considered, or, at least, not having discovered the method of reasoning from the happening of an event to the probability of one or another cause. The questions treated in the third chapter of this work could not therefore be attempted by them. Given an hypothesis presenting the necessity of one or another out of a certain, and not very large, number of consequences, they could determine the chance that any given one or other of those consequences should arrive; but given an event as having happened, and which might have been the consequence of either of several different causes, or explicable by either of several different hypotheses, they could not infer the probability with which the happening of the event should cause the different hypotheses to be viewed.

But, just as in natural philosophy the selection of an hypothesis by means of observed facts is always preliminary to any attempt at deductive discovery; so in the application of the notion of probability to the actual affairs of life, the process of reasoning from observed events to their most probable antecedents must go before the direct use of any such antecedent, cause, hypothesis, or whatever it may be correctly termed. These two obstacles, therefore, the mathematical difficulty, and the want of an inverse method, prevented the science from extending its views beyond problems of that simple nature which games of chance present.

If he ever checked the reference to his quotation, therefore, Chrystal was scarcely playing fair. His case as well as Venn's illustrates the truth that the best causes tend to attract to their support the worst arguments, which seems to be equally true in the intellectual and in the moral sense.

3.1 Isaac Todhunter

In his *History of the Mathematical Theory of Probability* (1865) Todhunter devotes the fourteenth chapter to Bayes. The brief opening paragraph states, no doubt correctly, what was regarded as common ground at this period (p. 294).

"539. The name of Bayes is associated with one of the most important parts of our subject, namely, the method of estimating the probabilities of the causes by which an observed event may have been produced. As we shall see, Bayes commenced the investigation, and Laplace developed it and enunciated the general principle in the form which it has since retained".

Little fault need be found with the first sentence, save that it is unnecessarily vague. Bayes had proposed a method of *calculating*, with mathematical accuracy, and without uncertainty, the probability that the unknown probability of an observable event should lie between definite limits.

"541. Bayes proposes to establish the following theorem:". Todhunter states the conclusion unconditionally much as in expression (5) p. 12. However, he adds the reservation "Moreover we shall see that there is an important condition implied which we have omitted in the above enunciation, for the sake of brevity: we shall return to this point in Art. 552".

"552. It must be observed with respect to the result of Art. 549, that in Bayes' own problem we *know* that *a priori* any position of EF between AB and CD is equally likely; or at least we know what amount of assumption is involved in this supposition. In the applications which have been made of Bayes' theorem, and of such results as that which we have taken from Laplace in Art. 551, there has however often been no adequate ground for such knowledge or assumption".

Todhunter thus dissociates himself from the Laplacian generalization, and stresses the factual rather than the axiomatic basis proposed for Bayes' knowledge of probabilities *a priori*. It is indeed regrettable that, when he was on the point of having grasped this essential distinction, he should have wavered so far as to add "or at least we know what amount of assumption is involved in this supposition". One is inclined to ask:—Just how does this come to be known, and, if it does, what purpose is served by such knowledge? Bayes' knowledge is based on his imagined experiment with the billiard balls. It is essential for the demonstration of his Theorem. In chapter V Section 6 it will be shown that somewhat different experimental knowledge leads to a somewhat different conclusion.

It will be noticed that Todhunter, like Venn, whose book was published in the following year (1866), picks on the Rule of Succession as particularly open to criticism. However, the Rule of Succession follows as a direct deductive consequence from Bayes' expression,

$$\frac{(a+b+1)!}{a!b!} p^a q^b dp,$$

for the probability that the true value p should lie in the infinitesimal range dp, and is therefore true whenever Bayes' result is justified. It is the lack of experimental knowledge behind many seeming applications of Bayes' theorem, not the derivation from it of the Rule of Succession, that invalidates such arguments.

Near as he came to clarifying the situation, Todhunter's name cannot properly be added to those who finally succeeded in extricating the mathematical thought of the mid-nineteenth century from its bewildering difficulties. Though he knew of and expressed admiration for Boole's book of 11 years earlier, by slurring over the essential distinction which Boole had striven to make clear, his influence in effect was to confirm the textbook expositions current in his own time.

4. The meaning of probability

Whatever view may be preferred on the controversial issues which the quotations set out above have been selected to illustrate, it is evident beyond question that highly competent, or even illustrious, mathematicians had formed upon them quite irreconcilable opinions; and this appearance of inability to find a common ground is not lessened by a perusal of what has been written in our own century.

Since there is no reason to doubt the purely mathematical ability of these writers, it is natural to suspect a semantic difficulty due to an imperfect analysis of words regarded as being too simple to be elucidated by further examination, such as the word "probability" itself. Of course, each writer has "defined" this word to his own satisfaction. Mathematical definition is, however, often no more than a succinct statement of the axioms to be applied when the word occurs in deductive mathematical reasoning, and may pay less attention than is needed to the conditions of the correct applicability of the term in the real world. It is these conditions of applicability which are properly the concern of those responsible for Applied Mathematics.

Indeed, I believe that a rather simple semantic confusion may be indicated as relevant to the issues discussed, as soon as consideration is given to the meaning that the word probability must have to anyone so much practically interested as is a gambler, who, for example, stands to gain or lose money, in the event of an ace being thrown with a single die. To such a man the information supplied by a familiar mathematical statement such as: "If a aces are thrown in n trials, the probability that the difference in absolute value between a/n and $1/6$ shall exceed any positive value ϵ, however small, shall tend to zero as the number n is increased indefinitely", will seem not merely remote, but also incomplete and lacking in definiteness in its application to the particular throw in which he is interested. Indeed, by itself it says nothing about that throw. It is obvious, moreover, that many subsets of future throws, which may include his own, can be shown to give probabilities, in this sense, either greater or less than 1/6. Before the limiting ratio of the whole set can be accepted as applicable to a particular throw, a second condition must be satisfied, namely that before the die is cast no such subset can be *recognized*. This is a necessary and sufficient condition for the applicability of the limiting ratio of the entire aggregate of possible future throws as the probability of any one particular throw. On this condition we may think of a particular throw, or of a succession of throws, as a *random* sample from the aggregate, which is in this sense subjectively homogeneous and without recognizable stratification.

Makers of the standard apparatus of games of chance, dice, cards, roulettes, etc., take great care to satisfy both the requirements of a sufficiently specific statement of what is meant by probability. If either the long-run frequencies were faulty, or, in particular, if there were any means of foreseeing, even to a limited extent, the outcome of their use in a particular case, the apparatus, or, perhaps, the method of using them, would be judged defective for the purpose for which they were made.

This fundamental requirement for the applicability to individual cases of the concept of classical probability shows clearly the role both of well specified ignorance and of specific knowledge in a typical probability statement. It has been often recognized that any probability statement, being a rigorous statement involving uncertainty, has less factual content than an assertion of certain fact would have, and at the same time has more factual content than a statement of complete ignorance. The *knowledge* required for such a statement refers to a well-defined aggregate, or population of possibilities within which the limiting frequency ratio must be exactly known. The necessary *ignorance* is specified by our inability to discriminate any of the different sub-aggregates having different frequency ratios, such as must always exist. Laplace's definition of probability, in which he actually speaks of "évènemens", is so worded that this necessary stipulation of ignorance in respect of particular events can be transferred to *hypotheses*, so as to imply in Boole's words "the assigning to different states of things of which we know nothing, and upon the very ground that we know nothing, equal degrees of probability". Has Laplace not in fact passed unawares from proposition (*a*) below to proposition (*b*)?

(*a*) A possible outcome must be assigned equal probabilities in different future throws, because we can draw no relevant distinction between these in advance.

(*b*) Hypotheses must be judged equally probable *a priori* if no relevant distinction can be drawn between them.

How extremely conservative was the tradition of mathematical teaching is shown by the slowness with which the opinions of Boole, Venn and Chrystal were appreciated. The reluctance naturally felt to abandoning a false start was certainly enhanced by the fact that, so far as the problem of scientific induction was concerned, nothing had been put forward to replace that which had been taken away. The gap seems to have been felt only subconsciously. In many cases it must have been clear that it was possible for data of great value to the formation of our scientific ideas to be presented, and yet for there to be no defensible basis, in the light of the criticisms which had been made, for the application of Bayes' theorem. Many mathematicians must have felt that with a proper restatement, the theorem, or one fulfilling the same purpose in inductive reasoning, could be set on its feet again. Indeed, the two leading statisticians in England at the beginning of the twentieth century, K. Pearson (1920)[12] and F. Y. Edgeworth (1908)[4] (p. 387) both put forward attempts, discordant indeed and both abortive, to justify the mode of reasoning in which no doubt each had been brought up, but which had since been discredited.

The reader of this preliminary chapter will have seized my meaning if he perceives that the different situations in which uncertain inferences may be attempted admit of logical distinctions which should guide our procedure. That it may be that the data are such as to allow us to apply Bayes' theorem, leading to statements of probability; or secondly, that we may be able validly to apply a test of significance to discredit a hypothesis the expectations from which are widely at variance with ascertained fact. If we use the term rejection for our attitude to such a hypothesis, it should be clearly understood that no irreversible decision has been taken; that, as rational beings, we are prepared to be convinced by future evidence that appearances were deceptive, and that in fact a very remarkable and exceptional coincidence had taken place. Such a test of significance does not authorize us to make any statement about the hypothesis in question in terms of mathematical probability, while, none the less, it does afford direct guidance as to what elements we may reasonably incorporate in any theories we may be attempting to form in explanation of objectively observable phenomena. Thirdly, the logical situation we are confronted with may admit of the consideration of a series, or, more usually, of a continuum of hypotheses, one of which *must* be true, and among which a selection may be made, and that selection justified, so far as may be, by statistical reasoning. The stasis or deadlock which had set in by the end of the century has been, I shall hope to show, in fact, released by the consideration of these diverse possibilities.

REFERENCES

1. T. Bayes (1763). An essay towards solving a problem in the doctrine of chances. *Phil. Trans. Roy. Soc.*, vol. 53, p. 370.

2. G. Boole (1854). *The Laws of Thought.* Macmillan and Co., London. Reprinted by Dover Publications, Inc., New York, 1951.

3. G. Chrystal (1886). *Algebra.* Adam and Charles Black, London.

4. F. Y. Edgeworth (1908). On the probable errors of frequency-constants. *J. Roy. Stat. Soc.*, vol. 71, pp. 381-397.

5 R. A. Fisher (1935-1966). *The Design of Experiments.* Oliver and Boyd, Edinburgh.

6. R. A. Fisher (1958). Mathematical probability in the natural sciences. *18th Int. Congr. of Pharm. Sciences.* Brussels, Sept. 1958.

R. A. Fisher (1962) Some examples of Bayes' method of the experimental determination of probabilities *a priori*. *J. Royal Stat. Soc., B*, vol. 24, pp. 118–124.

7. P.-S. marquis de Laplace (1812, 1820). *Théorie analytique des probabilités.* Paris. 1st ed., p. 178. 3rd ed. (1820), preface, p. iv.

8. J. Michell (1767). An inquiry into the probable parallax, and magnitude of the fixed stars, from the quantity of light which they afford us, and the particular circumstances of their situation.
Phil. Trans., vol. 57, pp. 234-264.

9. A. de Moivre (1718, 1738, 1756). *Doctrine of Chances.*

10. P. De Montmort (1708, 1714). Essai d'analyse sur les jeux de hazards.

11. A. De Morgan (1838). *An Essay on Probabilities and on their Application to Life Contingencies and Insurance Offices.*
Longman and Co., London.

12. K. Pearson (1920). The fundamental problem of practical statistics.
Biometrika, vol. 13, pp. 1-16.

13. *The Dictionary of National Biography.* Oxford University Press.

14. I. Todhunter (1865). *A History of the Mathematical Theory of Probability.*
Macmillan and Co., London.

15. J. Venn (1866, 1876, 1888). *The Logic of Chance.*
Macmillan and Co., London.
Reprinted by Chelsea Publishing Co., New York, 1962.

CHAPTER III

FORMS OF QUANTITATIVE INFERENCE

1. The simple test of significance

While, as Bayes perceived, the concept of Mathematical Probability affords a means, in some cases, of expressing inferences from observational data, involving a degree of uncertainty, and of expressing them rigorously, in that the nature and extent of the uncertainty is specified with exactitude, yet it is by no means axiomatic that the appropriate inferences, though in all cases involving uncertainty, should always be rigorously expressible in terms of this same concept. Although this belief seems to have been unquestioned over the period of 150 years covered by the discussion of Chapter II, familiarity with the actual use made of statistical methods in the experimental sciences shows that in the vast majority of cases the work is completed without any statement of mathematical probability being made about the hypothesis or hypotheses under consideration. The simple rejection of a hypothesis, at an assigned level of significance, is of this kind, and is often all that is needed, and all that is proper, for the consideration of a hypothesis in relation to the body of experimental data available. It is therefore desirable to examine the logical nature of this sort of uncertain inferences.

The example chosen by Boole of Michell's calculation with respect to the Pleiades will serve as an illustration. He demonstrates that Bayes' method can be applied to this case only by assuming arbitrary values not provided by the data, and therefore that no probability *a posteriori* can be assigned to the hypothesis that the stars, down to the sixth magnitude, are distributed at random over the celestial sphere. He does not emphasize that nevertheless Michell had by his calculations presented a strong reason for rejecting this hypothesis, or attempt to exhibit just how such a rational inference should be correctly stated.

Michell supposed that there were in all 1500 stars of the required magnitude and sought to calculate the probability, on the hypothesis that they are individually distributed at random, that any one of them should have five neighbours within a distance of a minutes of arc from it. I find the details of Michell's calculation obscure, and suggest the following argument.

The fraction of the celestial sphere within a circle of radius a minutes is, to a satisfactory approximation,

$$p = \left(\frac{a}{6875 \cdot 5}\right)^2 \qquad (14)$$

in which the denominator of the fraction within brackets is the number of minutes in two radians. So, if a is 49, the number of minutes from Maia to its fifth nearest neighbour, Atlas, we have

$$p = \frac{1}{(140 \cdot 316)^2} = \frac{1}{19689}. \qquad (15)$$

Out of 1499 stars other than Maia of the requisite magnitude the expected number within this distance is therefore

$$m = \frac{1499}{19689} = \frac{1}{13 \cdot 1345} = \cdot 07613. \qquad (16)$$

The frequency with which 5 stars should fall

within the prescribed area is then given approximately by the term of the Poisson series

$$e^{-m}\frac{m^5}{5!}, \tag{17}$$

or, about 1 in 50,000,000, the probabilities of having 6 or more close neighbours adding very little to this frequency. Since 1500 stars have each this probability of being the centre of such a close cluster of 6, although these probabilities are not strictly independent, the probability that among them any one fulfils the condition cannot be far from and certainly cannot exceed 30 in a million, or 1 in 33,000. Michell arrived at a chance of only 1 in 500,000 but the higher probability obtained by the calculations indicated above is amply low enough to exclude at a high level of significance any theory involving a random distribution.

The force with which such a conclusion is supported is logically that of the simple disjunction: *Either* an exceptionally rare chance has occurred, *or* the theory of random distribution is not true.

In view of the efforts which have been made to force a frequency interpretation on to such a disjunction, it is to be noted that the mental reluctance to accept an event intrinsically improbable would still be felt if, for example, a *datum* were added to Michell's problem to the effect that it was a million to one *a priori* that the stars should be scattered at random. We need not consider what such a statement of probability *a priori* could possibly mean in the astronomical problem; all that is needed is that if this datum were introduced into the calculation, then, in view of the observations, a probability statement could be inferred *a posteriori*, to the effect that the odds were 30 to 1 that the stars really had been scattered at random. The inherent improbability of what has been observed being observable on this view still remains in our minds, and no explanation has been given of it. It has been overweighted, not neutralized, by the even greater supposed improbability of the universe chosen for examination being of the supposedly exceptional kind in which the stars are *not* distributed at random. The observer is thus not left at all in the same state of mind as if the stars had actually displayed no evidence against a random arrangement, although he would have been forced logically to admit that (so far as statements in terms of probability went) such a theory was probably true, and that the remarkable features that had attracted his attention were, incredible as it might seem, wholly fortuitous.

The example shows that the resistance felt by the normal mind to accepting a story intrinsically too improbable is not capable of finding expression in any calculation of probability *a posteriori*. The variety of ways in which this resistance does express itself very well exhibits its reality. Common reactions are:

(*a*) The whole thing is a fabrication.

(*b*) There is no sufficient reason to think that the facts were observed and put on record accurately.

(*c*) There has been exaggeration, and the omission of circumstances that would help to explain what is claimed.

(*d*) Some occult cause, beyond our present understanding, must be invoked.

In the studies known as parapsychology enormous

odds are often claimed, evidently with a view to raising the resistance felt to accepting what is intrinsically improbable to such a pitch that conclusion (*d*), although itself repugnant, shall be accepted in preference. The incredulous, however, tend to prefer explanations of types (*a*), (*b*) or (*c*) either to accepting such a claim as, let us say, "precognition", or, what seems almost always to be the last choice, to the acceptance as genuine of a very rare contingency.

The fact, important for the understanding of logical situations of this kind, that reluctance to accept a hypothesis strongly contradicted by a test of significance is not removed, though it may be outweighed, by information *a priori*, is exhibited also by the consideration that if the proposed datum, "The odds are a million to one *a priori* that the stars should really be distributed singly and at random"—if this datum were considered as a *hypothesis*, it would be rejected at once by the observations at a level of significance almost as great as the hypothesis, "The stars are really distributed at random", was rejected in the first instance. Were such a conflict of evidence, as has here been imagined under discussion, not in a mathematical department, but in a scientific laboratory, it would, I suggest, be some prior assumption, corresponding to an axiom or a datum in a mathematical argument, that would certainly be impugned.

The attempts that have been made to explain the cogency of tests of significance in scientific research, by reference to supposed frequencies of possible statements, based on them, being right or wrong, thus seem to miss the essential nature of such tests. A man who "rejects" a hypothesis provisionally, as a matter of habitual practice, when the significance is at the 1% level or higher, will certainly be mistaken in not more than 1% of such decisions. For when the hypothesis is correct he will be mistaken in just 1% of these cases, and when it is incorrect he will never be mistaken in rejection. This inequality statement can therefore be made. However, the calculation is absurdly academic, for in fact no scientific worker has a fixed level of significance at which from year to year, and in all circumstances, he rejects hypotheses; he rather gives his mind to each particular case in the light of his evidence and his ideas. It should not be forgotten that the cases chosen for applying a test are manifestly a highly selected set, and that the conditions of selection cannot be specified even for a single worker; nor that in the argument used it would clearly be illegitimate for one to choose the actual level of significance indicated by a particular trial as though it were his lifelong habit to use just this level. Further, the calculation is based solely on a hypothesis, which, in the light of the evidence, is often not believed to be true at all, so that the actual probability of erroneous decision, supposing such a phrase to have any meaning, may be, for this reason only, much less than the frequency specifying the level of significance. A test of significance contains no criterion for "accepting" a hypothesis. According to circumstances it may or may not influence its acceptability.

On the whole the ideas (*a*) that a test of significance must be regarded as one of a series of similar tests applied to a succession of similar bodies of data, and (*b*) that the purpose of the test is to discriminate or "decide" between two or more hypotheses, have greatly obscured their understanding, when taken not as contingent possibilities but as elements essential to their logic. The appreciation of such more complex cases will be much aided by a clear view of the nature of a test of significance applied to a single hypothesis by a unique body of observations.

Though recognizable as a psychological condition of reluctance, or resistance to the acceptance of a proposition, the feeling induced by a test of significance has an objective basis in that the probability statement on which it is based is a fact communicable to, and verifiable by, other rational minds. The level of significance in such cases fulfils the conditions of a measure of the rational grounds for the disbelief it engenders. It is more primitive, or elemental than, and does not justify, any exact probability statement about the proposition.

When a prediction is made, having a known low degree of probability, such as that a particular throw with four dice shall show four sixes, an event known to have a mathematical probability, in the strict sense, of 1 in 1296, the same reluctance will be felt towards accepting this assertion, and for just the same reason, indeed, that a similar reluctance is shown to accepting a hypothesis rejected at this level of significance. There is the logical disjunction: Either an intrinsically improbable event will occur, or, the prediction will not be verified. The psychological resistance has been, I think wrongly, ascribed to the fact that the event in question *has*, in the proper sense of the Theory of Probability, the low probability assigned to it, rather than to the fact, very near in this case, that the correctness of the assertion would *entail* an event of this low probability. The probability statement is a sufficient, but not a necessary, condition for disbelief in this degree. Disbelief is equally justified when the probability is hypothetical. The difficulty of traditional forms of expression, in this as in other cases, flows from the assumption, too widely disseminated, that Mathematical Probability, being the first well-defined concept, and for a long while the only one available, for the expression of statements of uncertainty, must necessarily be by itself competent for the adequate specification of uncertainty, that is of the grounds for belief or disbelief, in all logical situations. The logical consequences of a statement of Mathematical Probability are clear and well-known. They allow of the calculation of long-run frequency-ratios, and therefore to the habitual gambler, of long-run policies in laying bets, in the hypothetical and restricted field of games of chance, played fairly with perfect apparatus; but these logical consequences may, as in the examples under discussion, be in themselves of little importance to the bearing of observable facts on the acceptibility of possible hypotheses.

In general, tests of significance are based on *hypothetical* probabilities calculated from their null hypotheses. They do not generally lead to any probability statements about the real world, but to a rational and well-defined measure of reluctance to the acceptance of the hypotheses they test.

Closely related to the assumption that all expressions of uncertain knowledge must have the same logical form, namely that of a statement of probability, is the assumption that all kinds of evidence used as data for such inferences have the same kind of logical consequences. Even a writer so shrewd as J. M. Keynes (1921) has exposed himself to this criticism. In an illuminating passage he objected to the traditional way of speaking, which he ascribed to "Laplace and his followers", of a probability as "unknown". "Do we mean unknown through lack of skill in arguing from given evidence, or unknown through lack of evidence? The first is alone admissible, for new evidence would give us a new probability, not a fuller knowledge of the old one"[8] (p. 31).

In this he must not, as might at first be thought, be taken to assert that in all cases a probability statement can be made, for he says later on the same page "We ought to say that in the case of some arguments a relation of probability does not exist, and not that it is unknown". This admission greatly reduces the force of his objection, for if, at one stage, a probability does not exist, but with further observations it becomes possible to assert a definite probability, it is not a misuse of words to say that it was at first unknown, and has later been ascertained. However, for the cases of which he was thinking, I have the same preference as Keynes, and welcome his statement that in some cases no probability exists. In other cases which go back historically so far as Bayes, we undoubtedly have, or can logically conceive of having, partial knowledge of a probability, so that probability statements can be made about its value, and it would be stretching language unprofitably to say that in such cases the probability only partially exists.

Keynes' difficulty arises, I think, partly from a desire to use the word probability primarily to refer to the truth of propositions, and only to the occurrence of events in the sense that the probability of an event can be identified with the probability of the truth of the proposition that the event shall occur. In view of this identification there is no occasion to be fastidious, and certainly events, such as the disintegration within a determinate time of an atom of a radio-active element may be reasonably thought of as having an objective probability, independent of the state of our evidence, and this may be unknown, or known with a limited and determinate precision. The kind of evidence of which Keynes is thinking is the kind which, as it is increased indefinitely, would lead the probability inferred to tend to 0 or 1, that is to a statement without uncertainty. He does not notice that other kinds of evidence may lead to the estimation of an objective probability with increasingly high precision, so as to tend in the limit to an exact knowledge of its value; or, that those other kinds of evidence which can lead to no statement of probability whatever, may be of direct inferential value.

2. More general hypotheses

Scientific hypotheses usually differ from the simple hypothesis (the random distribution of the stars) considered by Michell, in that they allow of one or more parameters, or adjustable "constants", any value of which, or, any value within assigned limits, is consistent with the hypothesis. With respect to such hypotheses tests of significance may be applied in two ways. In the first place, a test of significance may be developed capable of rejecting the hypothesis as a whole, if any relevant feature of the observational record can be shown to depend on a contingency which is sufficiently rare whatever may be the value of the parameter. Secondly, if no such feature obtrudes itself, or, if any such as can be found is far-fetched and artificial, so that the general hypothesis is judged provisionally acceptable, the question arises of the Estimation of the parameter's value. In some such cases, but not all, as will be seen, inferences can be drawn assigning a calculable mathematical probability to any assertion to the effect that the parameter lies between assigned limits.

In choosing the grounds upon which a general hypothesis should be rejected, personal judgement may and should properly be exercised. The experimenter will rightly consider all points on which, in the light of current knowledge, the hypothesis may be imperfectly accurate, and will select tests, so far as possible, sensitive to these possible faults, rather than to others. Sometimes also he may make a comprehensive test, such as Pearson's test of Goodness of Fit, applied to observed frequencies, which though strictly speaking approximate, and inapplicable to small frequencies, is, albeit not specifically sensitive to particular faults, yet, when the frequencies are not too small, capable of detecting sufficiently pronounced discrepancies of any kind.

In experimental breeding the data available for theoretical discussion are the frequencies observable when a progeny, or group of progenies, is classified into a number of classes (phenotypes) the members of which are distinguishable by inspection, or by other tests. Thus in an intercross between organisms each heterozygous in respect of two genetic factors showing complete dominance, the expectations out of n observed will be

$$\frac{n}{16}(9, 3, 3, 1), \tag{18}$$

if the factors are unlinked, but in general

$$\frac{n}{4}(2+\theta,\ 1-\theta,\ 1-\theta,\ \theta)\,, \tag{19}$$

where θ is a parameter, depending on the closeness of linkage, any value of which between 0 and 1 would be intelligible.

In the general χ^2 method, if a_1, a_2, a_3, a_4 are frequencies observed corresponding with frequencies m_1 to m_4 expected, the measure of discrepancy is

$$\chi^2 = \overset{4}{\underset{1}{S}}\left\{\frac{(a-m)^2}{m}\right\} = \overset{4}{\underset{1}{S}}\left(\frac{a^2}{m}\right)-n\,, \tag{20}$$

and the probability with which the value of χ^2 observed will be exceeded by chance for three degrees of freedom can be found from the well-known tables. When the expectation is calculated for factors inherited independently, the χ^2 value is divisible into three positive and independent parts.

$$\left.\begin{aligned} &\frac{1}{3n}(a_1+a_2-3a_3-3a_4)^2\,, \\ &\frac{1}{3n}(a_1-3a_2+a_3-3a_4)^2\,, \\ \text{and}\qquad &\frac{1}{9n}(a_1-3a_2-3a_3+9a_4)^2\,, \end{aligned}\right\} \tag{21}$$

each of which is distributed, on the hypothesis tested, as χ^2 for one degree of freedom, that is, as the square of a normal deviate having unit variance. The last of these three parts is specifically sensitive to a disturbance of the expectations of the kind due to linkage; the first two are sensitive to disturbances of the single-factor ratios, and if the only disturbance really present is a small effect of linkage, the test using three degrees of freedom including them also, will be less sensitive to the linkage effect than that based on only one.

To test the general hypothesis the usual procedure for large samples is to estimate the value of θ, by an efficient estimate, to find the corresponding expectations, and to ascertain the level of significance of the χ^2 value obtained, using the Table at two degrees of freedom. A low probability indicates that the general hypothesis is to be rejected at the level of significance found; in other words, there is no value of θ for which the hypothesis is acceptable. We do not simply use the sum of the components of χ^2 for the two single-factor ratios in this test, for if linkage were present these two components would no longer be independent, and to combine the evidence of the two requires an estimate of the parameter. This problem has been discussed in detail in *Statistical Methods*, section 55.[3]

Although in testing the acceptability of a hypothesis in general, no reference to the theory of estimation is required, at least in this important class of cases, beyond the stipulation that expectations can be fitted efficiently to the data, yet when the general hypothesis is found to be acceptable, and accepting it as true, we proceed to the next step of discussing the bearing of the observational record upon the problem of discriminating among the various possible values of the parameter, we are discussing the theory of estimation itself. In this theory a case of peculiar simplicity arises when an estimate exists which, perhaps in conjunction with ancillary statistics, subsumes the whole of the information, relevant to the parameter, supplied by the observational record. Such estimates are termed exhaustive, and their special property may be expressed in various ways: (i) that, given the exhaustive statistic, every other estimate possible has a sampling distribution completely independent of the parameter to be estimated, and (ii) that the Likelihood Function of the parameter inferred from the sampling distribution of the exhaustive estimate and its ancillaries is exactly the same as that inferred from the original observations. In fact for all purposes of inference an exhaustive statistic, in association perhaps with certain ancillary values, which themselves have distributions independent of the parametric value, can replace the entire observational record from which it was calculated.

Exhaustive estimates do not always exist; it is therefore important to note that their existence is the first requirement of the mode of inductive reasoning to be developed below as the Fiducial Argument. A second condition to be noted is that the observations should not be discontinuous, as are frequencies, but should be measurements sufficiently accurate to be regarded without significant error as observed values of continuous variates, so that the statistics calculable from them shall have continuous distributions. In making these distinctions, I do not wish to deny either that measurements, however accurate, are in the strictly mathematical sense discontinuous, or that counts may be of numbers so large that they could without sensible error be treated as measurements of continuous variables. It is only necessary to point out that cases commonly occur to which this distinction is relevant. In the same way, it may be said that it can always be imagined that statistical samples are made so large that the distinction between exhaustive and other efficient estimates shall become unimportant. All that is needed is to recognize that samples are not always so large as this, and that in such cases the logical consequences which flow from this distinction are not irrelevant.

3. The fiducial argument *

The term fiducial has been introduced to distinguish a particular form of inductive reasoning from that of Bayes, which for contrast may be termed the Bayesian argument. The distinction was needed because, like the method of Bayes, it leads to probability statements applicable in the light of the observations to an unknown parameter. Whereas, however, the argument of Bayes requires a distribution *a priori* involving probability statements of the same logical form as those finally obtained *a posteriori*, the application of the fiducial argument can only be made in the absence of such information *a priori*. In the Bayesian argument the observations are used to convert a random variable having a well-defined distribution *a priori* to a random variable having an equally well-defined distribution *a posteriori*; and it is well known that, if the observations are increased in number their importance grows relatively to that of the information supplied *a priori*, so that the latter becomes less and less influential upon the conclusions. By contrast, the fiducial argument uses the observations only to change the logical status of the parameter from one in which nothing is known

* Probability statements derived by arguments of the fiducial type have often been called statements of "fiducial probability". This usage is a convenient one, so long as it is recognized that the concept of probability involved is entirely identical with the classical probability of the early writers, such as Bayes. It is only the mode of derivation which was unknown to them.

of it, and no probability statement about it can be made, to the status of a random variable having a well-defined distribution.

If direct and exact observations could be made on the parameter itself, a similar change of logical status would be effected by the observation of its value, from one in which it was wholly unknown, or had perhaps a known frequency distribution, to one in which it could be assigned a definite value. It is, therefore, perhaps not surprising that similar exact observations, though not on the parameter itself yet on variates having distributions known in terms of the parameter, should be able in favourable cases to effect, at a lower level, a similar change of status.

As an example of the mode of reasoning, consider a radio-active source emitting particles with unknown frequency at instants completely independent of each other. The interval of time between two successive emissions will then be distributed at random in the exponential distribution

$$df = \theta e^{-\theta x}\, dx\,, \tag{22}$$

in which θ is the unknown average number of emissions in each time unit. We may conceive such time intervals to be accurately measurable, and that a record of n of them has shown intervals

$$x_1\,,\ x_2\,,\ x_3\,,\ \ldots,\ x_n\,.$$

We suppose that these conform sufficiently well to expectations based on the estimate

$$\left.\begin{aligned} \theta &= T\,, \\ \text{where} \qquad T &= \frac{n}{X}\,, \end{aligned}\right\} \tag{23}$$

and X stands for the sum of the times observed, for it to be agreed that the general hypothesis is acceptable, and that what remains is only to make such statements about the value of θ as the data allow.

From the original data, the n observed time intervals being independent, the Mathematical Likelihood of any value θ which the parameter may take is seen to be proportional to

$$\theta^n e^{-\theta X},$$

which is maximized at the value

$$\hat{\theta} = \frac{n}{X};$$

so that the estimate T chosen above is an estimate of maximum likelihood. It is also a Sufficient Estimate, that is to say an exhaustive estimate without ancillary statistics, for the sampling distribution of X is seen easily to be

$$df = \theta^n e^{-\theta X} \cdot \frac{X^{n-1}}{(n-1)!}\, dX\,, \tag{24}$$

giving exactly the same likelihood function for θ as was given by the original data. The distribution of X also is continuous over all positive values, uniformly for all values of θ.

This distribution of X for any given θ is, in fact, equivalent to the distribution of the quantity χ^2 for $2n$ degrees of freedom, if χ^2 is equated to

$$\chi^2 = 2\theta X = 2n\frac{\theta}{T}. \tag{25}$$

In this case the χ^2 distribution is exact, and not only approximate as is Pearson's measure of discrepancy for frequencies; consequently if we choose any probability P, and write

$$\chi^2_{2n}(P)$$

for that value which is exceeded for $2n$ degrees of freedom with frequency P, a value which is calculable with exactitude for all n and P, it appears that the statement

$$\theta > \frac{T}{2n} \chi^2_{2n}(P) \tag{26}$$

is verified with the frequency P, for all values of P chosen, and therefore that we have derived formally a frequency distribution of the unknown parameter θ appropriate to the observations available.

The applicability of the probability distribution to the particular unknown value of θ sought by an experimenter, without knowledge *a priori*, on the basis of the particular value of T given by his experiment, has been disputed, and certainly deserves to be examined, especially as in the first case in which I exhibited this form of argument, namely to the correlation coefficient (1930),[4] though the example was appropriate, my explanation left a good deal to be desired.

The reasoning developed so far has been entirely deductive; the example was chosen, however, to bring out some necessary characteristics of inductive reasoning. The probability statement first developed above (24) had as reference set all the values of X, and therefore of T, which might have occurred in unselected samples for a particular value of θ. It has, however, been proved for all values of θ, and so is applicable to the enlarged reference set of all pairs of values (T, θ) obtained from all values of θ. The particular pair of values of θ and T appropriate to a particular experimenter certainly belongs to this enlarged set, and within this set the proportion of cases satisfying the inequality (26)

$$\theta > \frac{T}{2n} \chi^2_{2n}(P)$$

is certainly equal to the chosen probability P. It might, however, have been true, as in the case of a gambler throwing a single die, discussed on page 32, that in some recognizable sub-set, to which his case belongs, the proportion of cases in which the inequality was satisfied should have some value other than P. It is the stipulated absence of knowledge *a priori* of the distribution of θ, together with the exhaustive character of the statistic T, that makes the recognition of any such subset impossible, and so guarantees that in his particular case, as in the case of a single particular throw contemplated by the gambler, the general probability is applicable.

Had knowledge *a priori* been available, the argument developed above would have been precluded by the consideration that some of the relevant data had been omitted. For, although in the deduction of statements of certainty it is legitimate to draw inferences from some of the axioms available while ignoring others, or, in other words to base a valid argument on a chosen subset only of the available axioms, no such liberty can be taken with statements of uncertainty, where it is essential to take the whole of the data into account, though some part of it may be shown on examination to be irrelevant, and not to affect the result.

Again, had there been knowledge *a priori*, the argument of Bayes could have been developed, which would have utilized all the data, and which would in general have led to a distribution *a posteriori* different from that to which the fiducial argument leads. Bayes' method in fact calculates the distribution of θ in a particular subset of pairs of values (T, θ), defined by T, and to which therefore the observation belongs. Consequently, it is essential to introduce the absence of knowledge *a priori* as a distinctive datum in order to demonstrate completely the applicability of the fiducial method of reasoning to the particular real and experimental cases for which it was developed. This last point I failed to perceive when, in 1930, I first put forward the fiducial argument for calculating probabilities. For a time this led me to think that there was a difference in logical content between probability statements derived by different methods of reasoning. There are in reality no grounds for any such distinction.

Various writers, including Sir Harold Jeffreys[7] and A. Kolmogorov,[9] recognizing the rational cogency of the fiducial form of argument, and the difficulty of rendering it coherent with the customary forms of statement used in mathematical probability, have proposed the introduction of new axioms to bridge what was felt to be a gap. The treatment in this book involves no new axiom; it does, however, rely on a property inherent in the semantics of the word "probability", though not required explicitly so long as the applicability in the real world of the logical relationship denoted is not in question. Purely abstract studies of the formal mathematics of probability can, in fact, be developed without reference to this aspect of the word's meaning. It is not, of course, unusual that mathematical definition should have, as it often does, axiomatic implications. The distinction should be made, in this case, that the completion of the word's definition specifies the nature and extent, not of the knowledge but explicitly of the ignorance, required in the logical situation envisaged, and that so long as it is assumed, as in purely deductive reasoning is proper, that valid deductions can be drawn from every subset of the axiomatic material available, it can be argued, as by Venn, that "his ignorance affects himself only, and corresponds to no distinction in the things". Mathematical probability, however, as conceived by the early writers, was applicable to the real world, and to make it available not only in deductive, but also in inductive reasoning a more complete definition is required. The subject of a statement of probability must not only belong to a measurable set, of which a known fraction fulfils a certain condition, but every subset to which it belongs, and which is characterized by a different fraction, must be unrecognizable.

4. Accurate statements of precision

The possibility of making exact statements of probability about unknown constants of Nature supplies a need long felt of making a complete specification of the precision with which such constants are estimated. For example, if, in the case considered in the foregoing section, there had been 500 accurately measured time intervals, calculations based on the distribution of χ^2 for 1000 degrees of freedom would show that the probability was 25% of the true value lying below ·96957 of the estimate, and 25% of lying above 1·02988, times the same quantity. These values then bracket a central region of about 6% within which the true value will lie with a probability of just one half. They therefore fulfil the same function as the traditional "probable error" often given in respect of astronomical observations. The concept of the probable error indicates the desire felt for such probability statements, although the great complexity of the observational material to be reduced in many astronomical calculations has stood in the way of the refinement of the concept, and has indeed often introduced great difficulties in the way of obtaining a reliable figure. When, as in the example chosen, the data are simple and the meaning of the calculations completely clear, other relevant probability statements may be made with equal confidence and exactitude. For example, the probability is 5% each way of the true value lying outside the limiting ratios ·92731 and 1·07439, and it is only 1% of it lying below ·89819 and another 1% of lying above 1·10622, so that the odds are 49 to 1 that it should lie within these last limits. The fiducial distribution in this way comprises a complete set of probability statements appropriate to any chosen level of probability, or to any chosen limits. In such cases the precision of the estimate has been completely specified.

Had there been a smaller number of time-measurements, such as 15, the precision of the estimate would have been lower, but the fiducial limits and corresponding statements of probability would still have been exact. Compendiously the corresponding results may be stated as follows:

Excluding at each end	Lower limit	Upper limit
1%	·4984	1·6964
5%	·6164	1·4591
25%	·8159	1·1600

There is here a probability of over 1% that the true value should be less than half the estimate.

The objection has been raised that since any statement of probability to be objective must be verifiable as a prediction of frequency, the calculations set out above cannot lead to a true probability statement referring to a particular value of T, for the data do not provide the means of calculating this. This seems to assume that no valid probability statement can be made except by the use of Bayes' theorem. However, the aggregate of cases of which the particular experimental case is one, for which the relative frequency of satisfying the inequality statement is known to be P, and to which all values of T are admissible, could certainly be sampled indefinitely to demonstrate the correct frequency. In the absence of a prior distribution of population values there is no meaning to be attached to the demand for calculating the results of random sampling among *populations*, and it is just this absence which completes the demonstration that samples giving a particular value T, arising from a particular but unknown value of θ, do not constitute a distinguishable sub-aggregate to which a different probability should be assigned. Probabilities obtained by a fiducial argument are objectively verifiable in exactly the same sense and in exactly the same way as are the probabilities assigned in games of chance.

It has, as was shown in the previous chapter, been a hope or ambition among many mathematicians of the last two hundred years that the concept of Mathematical Probability should be found to be applicable, not only to idealized games of chance, but to practical affairs, and in particular to inferences in the Natural Sciences. The fiducial argument demonstrates at least one meaningful application beyond that for which it was originally defined, and without needing the knowledge *a priori* required for Bayes' method of reasoning.

It may be noticed in the example chosen above, and in other cases in which a probability distribution can be calculated by the fiducial argument, that the region containing, for example, the lowest 1% of the frequency distribution is exactly that comprising values of the parameter which would have been rejected as too low by a valid test of significance at the 1% level. The two concepts should none the less be distinguished, for valid tests of significance at all levels may exist without the possibility of deducing by an accurate argument, a probability distribution for the unknown parameter. The direct step from the test of significance to a probability distribution cannot be sustained, and this circumstance has been responsible for some misunderstanding, and confusion of the terminology.

5. Discontinuous observations

The data discussed by Bayes in which a successes have been observed out of $(a+b)$ trials are discontinuous in character, unlike all examples suitable for exhibiting the fiducial argument, in the simple form so far exhibited. As has been mentioned, as the frequencies counted are increased, the mathematical niceties necessary for the accurate treatment of small samples, and the corresponding logical distinctions, become unimportant in their effects, and may therefore be ignored in practice. Nevertheless, no apology is needed for examining, as will be done later in this section, what inferences can properly be drawn when such data are treated quite exactly.

If the frequencies observed are very large, for

example, if both a and b are counts of many millions or hundreds of millions of independent instances, the probability of success might, if only ordinary levels of precision were in view, be recognized to have effectively the status of a directly observable quantity namely,

$$p = a/(a+b), \tag{27}$$

in which we may ignore the logical fact that p is equated to an estimate affected by errors of random sampling. At least we may ignore this fact after having satisfied ourselves, by reference to a more exact treatment, that the precision is such that for the purposes for which we need it, the observed value is sufficiently accurate, with sufficiently high probability.

If the frequencies are of intermediate size, of the order, let us say, of 1000 or 10,000, we probably should not be willing to ignore the sampling error, but might be content, as in the case of the χ^2 test of Goodness of Fit, to ignore the discontinuity of the observations. Comparing the expectations for any assigned value of p, with the observations, we should have, with $a+b=N$,

Expected	Observed
pN	a
qN	b

leading to

$$\left.\begin{aligned} \frac{(qa-pb)^2}{pq(a+b)} &= \chi^2, \\ \text{or,}\quad \frac{qa-pb}{\sqrt{pq(a+b)}} &= \chi, \end{aligned}\right\} \tag{28}$$

as the test of significance for one degree of freedom. χ is then normally distributed with unit variance.

Choosing any appropriate significance level, such as 2%, the value of χ^2 is known, in this case to be 5·412, and substituting this value, we have a quadratic equation, giving the two values of p for which the deviation from the observations is exactly at the 2% level of significance. The probability that p should be less than the lower root of the equation is 1%, as is that of it exceeding the higher root. Similar values, excluding 5% at each end, could be obtained by taking 2·706 for χ^2. When the discontinuity of the observations can be ignored, the fiducial argument justifies statements of probability, concerning the unknown value p, which is thus estimated as a random variable with a precision defined by a consistent aggregate of such probability statements.

Since the fiducial argument in full strictness cannot be applied, owing to the actual discontinuity of the data, it would be improper to regard this distribution, though precise in form, as more than an asymptotic solution of Bayes' problem, when knowledge *a priori* is absent. As such, it is, however, relevant that the mean value of p for given observational frequencies a, b may be expressed as

$$\bar{p} = \frac{a}{N} + \frac{b-a}{2N}\left\{1 - e^{\frac{1}{2}N}\sqrt{N}\int_{\sqrt{N}}^{\infty} e^{-\frac{1}{2}u^2}\,du\right\}, \qquad (28.1)$$

or

$$\frac{a}{N} + \frac{b-a}{2N^2} - \frac{3(b-a)}{2N^3} + \ldots \; . \qquad (28.2)$$

Now the mean of Bayes' distribution *a posteriori* admits of the expansion

$$\frac{a+1}{N+2} = \frac{a}{N} + \frac{b-a}{N^2} - \frac{2(b-a)}{N^3} + \cdots \qquad (28.3)$$

whereas had Bayes' calculation been carried out with the probability element *a priori* given by (6)

$$\frac{1}{\pi\sqrt{pq}}\,dp\,,$$

the mean would have been

$$\bar{p}=\frac{a+\frac{1}{2}}{N+1}=\frac{a}{N}+\frac{b-a}{2N^2}-\frac{b-a}{2N^3}+\frac{b-a}{2N^4}-\ldots \qquad (28.4)$$

For asymptotic agreement with the fiducial distribution, so far as the term in N^{-2}, Bayes' postulated distribution *a priori* should have been that given by expression (6), derived from the angular transformation. To this extent a particular given distribution *a priori* may be nearly equivalent to complete ignorance *a priori*.

Equation (28.1) is not, indeed, exact, since in the third term of the expansion (28.2) the effects of the non-normality of the binomial distribution become appreciable. If allowance be made for these by the method of Cornish and Fisher[1] (1937), an expansion may be obtained of the parameter p in terms of a normal deviate x, of which the earlier terms, carried now a few steps further than in the first edition, and arranged, after the first three in tabular form, are

$$a/N-\sqrt{ab}\,x/N^{3/2}+(b-a)(2x^2+1)/6N^2$$

$$+\left\{\begin{matrix} x^3 & x & \\ -2 & -7 & N^2 \\ 26 & 34 & ab \end{matrix}\right\}\div 72N^{5/2}\sqrt{ab}$$

$$+(b-a)\left\{\begin{matrix} x^4 & x^2 & x^0 & \\ -12 & 17 & 19 & N^2 \\ -276 & -644 & -148 & ab \end{matrix}\right\}\div 3240N^3ab$$

$$+\left\{\begin{array}{rrrl} x^5 & x^3 & x & \\ -36 & 470 & 265 & N^4 \\ 936 & 5564 & 5188 & N^2ab \\ -11268 & -39712 & -23804 & a^2b^2 \end{array}\right\} \div 48\times 3240\, N^{7/2}(ab)^{3/2}$$

$$+(b-a)\left\{\begin{array}{rrrrl} x^6 & x^4 & x^2 & x^0 & \\ 24 & 207 & -418 & -563 & N^4 \\ 240 & -261 & -2059 & -758 & N^2ab \\ 5784 & 28026 & 29882 & 3400 & a^2b^2 \end{array}\right\} \div 126\times 3240 N^4a^2b^2$$

$$+\left\{\begin{array}{rrl} x^7 & x^5 & \\ 30024 & -17016 & \\ -51192 & -570492 & \\ -1,014120 & -9,896200 & \\ 12,139416 & 76,126776 & \\ x^3 & x & \\ -396616 & 792709 & N^6 \\ -435102 & -3,185262 & N^4ab \\ -22,500120 & -10,404420 & N^2a^2b^2 \\ 122,206216 & 37,191656 & a^3b^3 \end{array}\right\} \div 112\times 3240^2\, N^{9/2}(ab)^{5/2}$$

From this expansion, substituting its average value for each power of x, it can be ascertained that the mean of the fiducial distribution is

$$\bar{p} = \frac{a}{N} + \frac{b-a}{2N^2} - \frac{b-a}{2N^3} + \frac{b-a}{2N^4} \cdots$$

agreeing so far as the fourth term with (28·4), the mean of the Bayesian distribution *a posteriori*, using the element *a priori*

$$\frac{1}{\pi\sqrt{pq}}\,dp\,.$$

I do not know for how many terms this agreement continues. It should, however, be noted that the variances of the two distributions are not the same. That for the fiducial distribution is the larger by $1/12N^2$, or more accurately by $1/12(N+1)(N+2)$.

An odd consequence of the analysis developed above is that the Rule of Succession derivable from the particular distribution of probability *a priori*

$$\frac{dp}{\pi\sqrt{pq}},$$

namely that the probability of success in the next trial is

$$\frac{a+1/2}{a+b+1}$$

is justifiable, at least to a remarkably high approximation, in the absence of any knowledge *a priori*; and this although the corresponding complete distribution *a posteriori* is not so justifiable.

For smaller numbers of independent observations than those for which the effects of discontinuity are negligible, a lower logical status is recognizable in which neither effectively exact definitive statements, nor statements in terms of Mathematical Probability, are possible, yet in which some information is available, and we are not in a state of complete ignorance. I have, indeed, recently[6] discussed a quasi-probabilistic specification of the inferences in this case, in terms which are, however, still too unfamiliar for inclusion here. Evidently, also, in such cases tests of significance are available. Thus, to take an example employed in illustrating the Table of z, in the Introduction to *Statistical Tables*,[5] if 3 successes have been observed out of 14 trials, then probabilities of success exceeding ·557 may be excluded by the consideration that for such values the total probability of observing 0, 1, 2 or 3 successes is only 1%, while values below ·0331 may be excluded on the ground that the total probability of values from 3 to 14 would similarly fall below 1%. In this way tests of significance are available capable of finding at any level of significance limits outside which all values of the parameter are to be rejected. These have been called "Confidence Limits", and though they fall short in logical content of the limits found by the fiducial argument, and with which they have often been confused, they do fulfil some of the desiderata of statistical inferences.

(i) With the aid of the Table of the z-distribution, or of other tables serving the same purpose of determining the partial sum of the terms of the binomial expansion,

$$(q+p)^n,$$

the Confidence Limits can be sufficiently easily calculated.

(ii) They serve to divide the range of possible values of the unknown into a series of zones of more or less acceptable values.

(iii) The results of the calculations are readily communicable, and the method is sufficiently known to be widely understood.

Nevertheless, from the point of view of making the most of limited data, and of drawing from them conclusions as strong as they can properly be made, the system of Confidence Limits seems to provide less than even this comparatively uninformative type of data would support.

It has been frequently stated, as though it were a

characteristic property of Confidence limits, that the interval between them will in repeated samples cover the true value with the exact frequency corresponding with the level of significance chosen. E.g. that in 98% of trials the true value would be found to lie between the two 1% points. This, if true, when exhaustive estimation has been used, would give them the force of a statement of probability. However, actually, the true value will lie between the assigned limits generally in more than 98% of such trials, and no exact statement of probability can be inferred. Exactly verifiable probability statements are not a characteristic of Confidence limits, as they are of the limits that can be assigned when the fiducial argument is available.

As the probability defining the level of significance is raised the width of the central zone still deemed acceptable is narrowed, but it is not closed until the probability level is raised to considerably over 50%. At the value of p suggested by the data, namely 3/14 (21·42857%), the probability of observing 3 or less is 64·832% and that of observing 3 or more is 60·402%. At $p = \cdot 22$, the probability of 3 or less is 62·807% and that of 3 or more is 62·394. This system of zones therefore closes at some value slightly greater than $p = \cdot 22$, and at this value the level of significance is over 62%. Neither value seems to have any sufficient inferential content to be worthy of record and report. The method of zoning by Confidence Intervals does not pick out as of any importance the value of p for which the Mathematical Likelihood is greatest, or, in other words, that which would have the highest probability of leading to the result observed.

6. Mathematical Likelihood

Objection has sometimes been made that the method of calculating Confidence Limits by setting an assigned value such as 1% on the frequency of observing 3 or less (or at the other end of observing 3 or more) is unrealistic in treating the values less than 3, which have not been observed, in exactly the same manner as the value 3, which is the one that has been observed. This feature is indeed not very defensible save as an approximation. It should be pointed out that when the probability of 3 or less is small, most of this small probability will be due to the case "exactly 3", and that the contribution of the other three cases is not very important, although it does increase or decrease with varying p at a relative rate different from the contribution of "exactly 3" itself. Similarly, at values of p at which "3 or more" has a low probability, a large part of this probability will be due to the particular case that has been observed, and the calculation will have been perhaps little influenced by the frequencies of the cases, included in the calculation, which have not in fact been observed.

It would, however, have been better to have compared the different possible values of p, in relation to the frequencies with which the actual values observed would have been produced by them, as is done by the Mathematical Likelihood, a function of the unknown parameter proportional to these frequencies, or in this case to

$$p^a(1-p)^b,$$

having its maximum value at

$$p=a/(a+b),$$

and therefore expressible in terms of its maximum, as

$$\left.\begin{array}{c}\left(\frac{p}{a}\right)^a\left(\frac{1-p}{b}\right)^b(a+b)^{a+b}, \\ \text{or} \quad \frac{(a+b)^{a+b}}{a^a b^b}\cdot p^a(1-p)^b.\end{array}\right\} \quad (29)$$

The Mathematical Likelihood assignable to every value of the unknown parameter p supplies a zoning of the admissible range of values more directly appropriate to the observations than that provided by the system of Confidence belts. Mathematical Likelihood is not, of course, to be confused with Mathematical Probability. It is, like Mathematical Probability, a well-defined quantitative feature of the logical situations in which it occurs, and like Mathematical Probability can serve in a well-defined sense as a "measure of rational belief"; but it is a quantity of a different kind from probability, and does not obey the laws of probability. Whereas such a phrase as "the probability of *A* or *B*" has a simple meaning, where *A* and *B* are mutually exclusive possibilities, the phrase "the likelihood of *A* or *B*" is more parallel with "the income of Peter or Paul"—you cannot know what it is until you know which is meant.

In relation to the logical situations so far discussed Mathematical Likelihood has already appeared as the factor, appropriate to each possible parametric value, by which each element of probability *a priori* is converted in Bayes' method to the corresponding element of probability *a posteriori*. It represents that part of Bayes' calculation provided by the data themselves. With regard to simple tests of significance, if such tests were performed on the same data against several mutually exclusive hypotheses, since the likelihood of any hypothesis is proportional to the probability, accepting that hypothesis as true, of such observations occurring as have been made, the greatest reluctance will be felt to accepting the least likely hypothesis, and the least reluctance to the most likely. Such comparisons can be made even though only relative values of the likelihood function are meaningful. The likelihood supplies a natural order of preference among the possibilities under consideration. It is not surprising, therefore, though independently demonstrable, that in the Theory of Estimation, all rational criteria of what is to be desired in an estimate converge on the particular value for which the likelihood is maximized. The Method of Maximum Likelihood is indeed much used and widely appreciated in the statistical literature, without, I fancy, so much appreciation of the significance of the system of likelihood values at other possible values of the parameter. In the theory of estimation[2] it has appeared that the whole of the information supplied by a sample, within the framework of a given sampling method, is comprised in the likelihood, as a function known for all possible values of the parameter.

The relation between probability and likelihood in the case in which probabilities are accessible by the fiducial argument, is intimate. If θ stand for an unknown parameter and T for a Sufficient, or Exhaustive, estimate of it, and if for all values of θ and T, the function,

$$P=F(T, \theta),$$

stand for the probability that a sample drawn from a population with parameter θ, shall yield a statistic less than T, both θ and T being continuous functions over the same range, and F monotonic for both variables, then the distribution of T for given θ has the frequency element,

$$\frac{\partial F}{\partial T} dT ,$$

so that the likelihood of any value θ is determined by the relation

$$e^{L(\theta)} \propto \frac{\partial F}{\partial T} ,$$

for the particular value of T observed, and for all values of θ. Concurrently, we have seen that the frequency distribution of θ to be inferred from the data has the frequency element

$$-\frac{\partial F}{\partial \theta} d\theta .$$

The likelihood function and the probability distribution thus supply complementary specifications of the same situation.

"Confidence Limits" and "Confidence Belts" were I think developed and advocated under the impression that in a wider class of cases they could provide information similar to that of the probability statements derived by the fiducial argument. It is clear, however, that no exact probability statements can be based upon them, and this seems now to be understood. They may be taken to supply statements of inequality. The tests of significance on which they are based are, of course, valid, but if these are used for zoning the possible values of the parameter, the zones they give do not assign exactly the same order of preference as that supplied by the likelihood function. This is due to the use of heterogeneous groups of possibilities of which some have been observed, and others have not, in making up the blocks for which the probabilities are calculated.

For all purposes, and more particularly for the *communication* of the relevant evidence supplied by a body of data, the values of the Mathematical Likelihood are better fitted to analyse, summarize, and communicate statistical evidence of types too weak to supply true probability statements; it is important that the likelihood always exists, and is directly calculable. It is usually convenient to tabulate its logarithm, since for independent bodies of data such as might be obtained by different investigators, the "combination of observations" requires only that the log likelihoods from different sources should be added.

In the case under discussion a simple graph of the values of the Mathematical Likelihood expressed as a percentage of its maximum, against the possible values of the parameter p, shows clearly enough what values of the parameter have likelihoods comparable with the maximum, and outside what limits the likelihood falls to levels at which the corresponding values of the parameter become implausible. In Fig. 1 the likelihood is plotted against p. If instead of p, a transformed value, such as the normal deviate commonly used in Biological Assay is employed, the curve is transformed using invariant ordinates, and not, as would be the case with a frequency curve, with invariant areas. This is shown in Fig. 2. The areas under these curves are irrelevant. In each diagram zones are indicated showing the limits within which the likelihood exceeds 1/2, 1/5, and 1/15 of the maximum. Values of the parameter outside the last

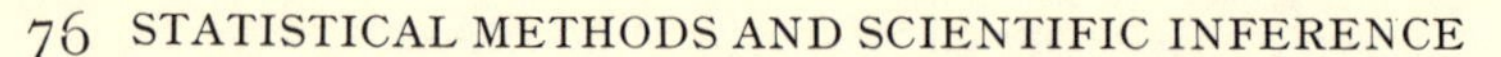

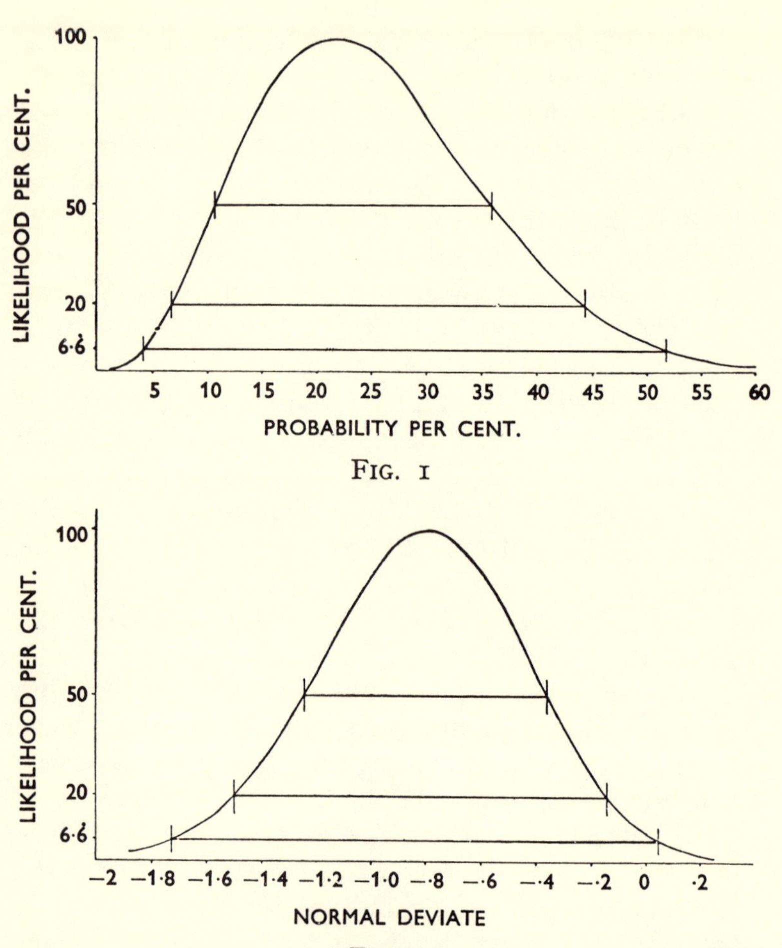

FIG. 1

FIG. 2

limits are obviously open to grave suspicion. The actual limits found are:

Likelihood ratio	Lower limit %	Upper limit %
50%	10·5889	35·9225
20%	6·6652	44·3301
6·6̇%	4·2001	51·6876

The simplicity of the data chosen for this example is an unessential accident. Very extensive observations may best be summarized in terms of the likelihood function calculable from them. A schedule of log likelihoods such as that shown below, in a form suitable for the information of a later worker, may be much more compact than the data from which it has been derived, and yet convey all that is needed from them:

p%	$-3 \log p$ $-11 \log (1-p)$	p%	$-3 \log p$ $-11 \log (1-p)$
3	4·714147	20	3·162920
4	4·388837	25	3·180506
5	4·148130	30	3·272558
6	3·961139	35	3·425749
8	3·689064	40	3·634156
10	3·503332	45	3·896373
12	3·373147	50	4·214420
14	3·282132	55	4·593574
17	3·198794	60	5·042886

Apart from the simple test of significance, therefore, there are to be recognized and distinguished, between the levels of certain knowledge and of total nescience, two well-defined levels of logical status for parameters lying on a continuum of possible values, namely that in which the probability is known for the parameter to lie between any assigned values, and that in which no probability statements being possible, or only statements of inequality, the Mathematical Likelihood of all possible values can be determined from the body of observations available.

By means of appropriate observations a quantity may conceivably pass discontinuously from one status to another implying fuller knowledge. Alternatively, the mere accumulation of data of the same kind on a sufficient scale may induce a kind of asymptotic approach to a higher status.

The implications of such a classificatory framework may be made clear by a greater variety of examples such as will be discussed in later chapters.

REFERENCES

1. E. A. Cornish and R. A. Fisher (1937). Moments and cumulants in the specification of distributions. *Rev. Inst. int. Statist.*, vol. 4, pp. 307-320.
2. R. A. Fisher (1925). Theory of statistical estimation. *Proc. Camb. Phil. Soc.*, vol. 22, pp. 700-725.
3. R. A. Fisher (1925-1958). *Statistical Method for Research Workers.* Oliver and Boyd, Edinburgh.
4. R. A. Fisher (1930). Inverse probability. *Proc. Camb. Phil. Soc.*, vol. 26, pp. 528-535.
5. R. A. Fisher and F. Yates (1938-1963). *Statistical Tables.* Oliver and Boyd, Edinburgh.
6. R. A. Fisher (1957) The underworld of probability. *Sankhya*, vol. 18, pp. 201-210.
7. H. Jeffreys (1940). Note on the Behrens-Fisher formula. *Ann. Eugen.*, vol. 10, pp. 48-51.
8. J. M. Keynes (1921). *A Treatise on Probability.* Macmillan and Co., London.
9. A. N. Kolmogorov (1942). The estimation of the mean and precision of a finite sample of observations. Section 5—Fisher's fiducial limits and fiducial probability. *Bull. Acad. Sci. U.S.S.R., Math. Ser.*, vol. 6, pp. 3-32.

CHAPTER IV

SOME MISAPPREHENSIONS ABOUT TESTS OF SIGNIFICANCE

1. Tests of significance and acceptance decisions

The common tests of significance, familiarly known as Pearson's χ^2 test of goodness of fit (1900),[22] "Student" 's *t*-test (1908),[24] the z (or F) test of the analysis of variance (1924),[4] and many others designed on the same principles, have come in the first two quarters of the twentieth century to play a rather central part in statistical analysis. In the day-to-day work of experimental research in the natural sciences, they are constantly in use to distinguish real effects of importance to a research programme from such apparent effects as might have appeared in consequence of errors of random sampling, or of uncontrolled variability, of any sort, in the physical or biological material under examination. They are used to recognize, among innumerable examples that could be given, the genuineness of a genetic linkage, the reality of the response to manurial treatment of a cultivated crop, the deterioration of a food product in storage, or the difference between machines in the frequency of defective parts produced by them. The conclusions drawn from such tests constitute the steps by which the research worker gains a better understanding of his experimental material, and of the problems which it presents.

It is noteworthy, too, that the men who felt the need for these tests, who first conceived them, or later made them mathematically precise, were all actively concerned with researches in the natural sciences. More recently, indeed, a considerable body of doctrine has attempted to explain, or rather to reinterpret, these tests on the basis of quite a different model, namely as means to making decisions in an acceptance procedure. The differences between these two situations seem to the author many and wide, and I do not think it would have been possible to overlook them had the authors of this reinterpretation had any real familiarity with work in the natural sciences, or consciousness of those features of an observational record which permit of an improved scientific understanding, such as are particularly in view in the design of experiments. The misapprehensions, indeed, appear to go deeper than would be expected from a mere transference of techniques from one field of study to another.

In various ways what are known as acceptance procedures are of great importance in the modern world. When a large concern such as the Royal Navy receives material from its makers, it is, I suppose, subjected to sufficiently careful inspection and testing to reduce the frequency of the acceptance of faulty or defective consignments. The instructions to the officers carrying out the tests must also, I conceive, be such as to keep low both the cost of testing and the frequency of the rejection of satisfactory lots. Much ingenuity and skill must be exercised in making an acceptance procedure a really effectual and economical one. It is not therefore at all in disdain of an artifice of proved value, in commerce and technology, that I shall emphasize some of the various ways in which this operation differs from that by which improved theoretical knowledge is sought in experimental research. This emphasis is primarily necessary because the needs and purposes of workers in the experimental sciences have been so badly misunderstood and misrepresented. It is, of course, also to be suspected that those authors, such as Neyman and Wald, who have treated these tests with little regard to their purpose in the natural sciences, may not have been more successful in the application of their ideas to the needs of acceptance procedures. It is, however, to the evident advantage of both kinds of application that the theories developed and taught to mathematicians should not confuse their several requirements.

In attempting to identify a test of significance as used in the natural sciences with a test for acceptance, one of the deepest dissimilarities lies in the population, or reference set, available for making statements of probability. Confusion under this head has on several occasions led to erroneous numerical values; for, where acceptance procedures are appropriate, the population of lots of one or more items, which could be chosen for examination, is unequivocally defined. The source of supply has an objective empirical reality. Whereas, the only populations that can be referred to in a test of significance have no objective reality, being exclusively the product of the statistician's imagination through the hypotheses which he has decided to test, or usually indeed of some specific aspect of these hypotheses. The demand was first made, I believe, in connection with Behrens' test of the significance of the difference between the means of two populations of unknown variances, that the level of significance should be determined by "repeated sampling from the same population", evidently with no clear realization that the population in question is hypothetical, that it could be defined in many ways, and that the first to come to mind may be quite misleading; or, that an understanding, of what the information is which the test is to supply, is needed before an appropriate population, if indeed we must express ourselves in this way, can be specified. This particular case will be examined more fully in a later section, after illustrating the more general effects of the confusion between the level of significance appropriately assigned to a specific test, with the frequency of occurrence of a specified type of decision.

2. "Student" 's Test

In the test of significance due to "Student" (W. S. Gossett), and generally known as the *t*-test, the data are taken to consist of N values of a single observable variate x, which are to be interpreted on the hypothesis that they are independent values of a variate normally distributed, with both mean and variance unknown, and of which no probability statements *a priori* are available. The two statistics

$$\left.\begin{aligned} \bar{x} &= \frac{1}{N} S(x) \\ s^2 &= \frac{1}{N-1} S(x-\bar{x})^2 \end{aligned}\right\} \qquad (30)$$

are known to be jointly Sufficient for the estimation of the true mean, μ, and the true variance, σ^2. Their sampling distributions in random samples of N may be expressed exactly in terms of the population parameters. Thus, as was shown by Gauss, the mean $\bar{x}$ has a Normal distribution about the true mean, μ, with the population variance divided by N. I.e. the frequency element is

$$\sqrt{\frac{N}{2\pi\sigma^2}}\, e^{-N(\bar{x}-\mu)^2/2\sigma^2}\, d\bar{x}\,, \tag{31}$$

while the distribution of s^2, due first I believe to Helmert[17] (1875) is independent of μ, and of $\bar{x}$, though involving σ^2. If u stand for

$$u = \tfrac{1}{2}\,\frac{(N-1)s^2}{\sigma^2}\,, \tag{32}$$

it takes the form

$$\frac{1}{\frac{N-3}{2}!}\, u^{\frac{N-3}{2}}\, e^{-u}\, du\,, \tag{33}$$

an Eulerian distribution, identifiable with that of χ^2 for $N-1$ degrees of freedom, by the equivalence

$$u = \tfrac{1}{2}\chi^2\,. \tag{34}$$

On the basis of these distributions "Student" set himself to ascertain the exact sampling distribution of the ratio between the sampling error of the mean, and its standard error *as estimated*, namely

$$t = \frac{\sqrt{N}(\bar{x}-\mu)}{s} \tag{35}$$

when $\bar{x}$ and s are calculated from a finite sample, and he was successful in establishing the frequency element

$$\frac{1}{\sqrt{\pi(N-1)}}\,\frac{\frac{N-2}{2}!}{\frac{N-3}{2}!}\cdot\frac{dt}{\left(1+\frac{t^2}{N-1}\right)^{N/2}}\,, \tag{36}$$

a distribution depending on N only, independent of both parameters of the Normal distribution sampled, and which can, therefore, be tabulated so as to give the value of t exceeded in absolute value, with any given probability, and for any number, $n=(N-1)$, of the degrees of freedom.

It will be recognized that "Student" 's distribution allows of induction of the fiducial type, for the inequality

$$\mu<\bar{x}-\frac{1}{\sqrt{N}}ts\,, \tag{37}$$

will be satisfied with just half the probability for which t is tabulated, if t is positive, and with the complement of this value if t is negative. The reference set for which this probability statement holds is that of the values of μ, $\bar{x}$ and s corresponding to the same sample, for all samples of a given size of all normal populations. Since $\bar{x}$ and s are jointly Sufficient for estimation, and knowledge of μ and σ *a priori* is absent, there is no possibility of recognizing any sub-set of cases, within the general set, for which any different value of the probability should hold. The unknown parameter μ has therefore a frequency distribution *a posteriori* defined by "Student" 's distribution.

Although his was the first exact test of significance, characteristic of the modern period, "Student" did not go so far as to claim that he was introducing a new mode of reasoning, and perhaps would have been unwilling to believe it had he been told so; for he was only applying his own good sense to a logical situation with which he was quite familiar. He was usually content to leave the inference in the form of a test of significance, namely that, on the hypothesis, for example, that the true mean of the population was zero, the value of t observed, or a greater value, would have occurred in less than 1% of trials, and was therefore significant at, for example, the 1% level. It does appear, however, at one point, that he certainly was thinking of the unknown mean as having, in the light of the observations, a definable frequency distribution, for in the extreme case of only two observations, x_1 and x_2, when the distribution of t reduces to Cauchy's distribution,

$$\frac{1}{\pi} \cdot \frac{dt}{1+t^2}, \tag{38}$$

and

$$s^2 = \tfrac{1}{2}|x_1 - x_2|^2, \tag{39}$$

so that

$$t = \frac{2(\bar{x} - \mu)}{|x_1 - x_2|}, \tag{40}$$

or exactly $+1$ or -1 if μ takes one or other of the two observed values, he does point out that these are the quartiles of Cauchy's distribution, or the points cutting off a quarter of the area on each side.

Thus by integration of his general formula "Student" had shown, in this case, that the mean of the population sampled had exactly the probability of one half of lying between the two values observed; this conclusion is notable as illustrating a mode of inference entirely independent of the form of the distribution, save that it be continuous, namely that as each observation has a half chance of being above and a half chance of being below the median, and as these chances are independent, it could have been demonstrated that the median should lie between the two observations, these being our sole source of information about its position, with probability exactly one half. This is a typically fiducial argument, which would have been vitiated by the existence of information *a priori*. "Student", indeed, guards himself against this possibility by stipulating that his sample "itself affords the only indication of the variability". I take this to make clear also that it is not one of an objective series of similar samples from the same population existing in reality, though it can be regarded by an act of imagination as one of a hypothetical reference set.

This case is more complex than that dealing with radioactive emission discussed in the last Chapter (Section 3), for here the simultaneous estimation of two parameters is required. It is more simple than will generally be the case in statistical work, for in this case no characteristic of the sample (i.e. of the whole body of observations available) can be found to define a subset to which our sample belongs, and which might exhibit a different, and more relevant, frequency distribution. It is this simplicity that has deceived those writers who have considered this one alone of the practically useful tests of significance into ignoring such subsets, or thinking that when such subsets are available their existence can be ignored; and that a mere consideration of repeated sampling, in one of the many forms this may take, is sufficient to specify the level of significance appropriate.

3. The case of linear regression

A case which illustrates well how misleading the advice is to base the calculations on repeated sampling from the same population, if such advice were taken literally, is that of data suitable for the estimation of a coefficient of linear regression. Suppose we have N pairs of values (x, y) constituting the numerical data, and suppose also that it is given that, for each value of x, the values of y are normally distributed; that the mean y for any value of x is a linear function of that variable,

$$E_x(y) = Y_x = \alpha + \beta x\,, \qquad (41)$$

and that the variance of each distribution, though unknown numerically, is known to be the same for all values of x; i.e.

$$E_x(y - Y_x)^2 = \sigma^2\,. \qquad (42)$$

The distribution of x may also be given, with or without parameters, but this information is, as will be seen, irrelevant.

At least since the time of Gauss it has been known that if A, B, C stand for the sums of squares and products

$$S(x-\bar{x})^2,\ S(x-\bar{x})(y-\bar{y}),\ S(y-\bar{y})^2\,, \qquad (43)$$

then the best estimate of β, the slope of the regression line, is

$$b = \frac{B}{A}\,. \qquad (44)$$

The best estimate of σ^2 is

$$s^2 = \frac{1}{N-2}\left(C - \frac{B^2}{A}\right), \qquad (45)$$

and, for samples having the same fixed value for A, the estimates b will be normally distributed about the true value, β, with sampling variance

$$V(b) = \frac{\sigma^2}{A}\,. \qquad (46)$$

We may therefore identify "Student" 's t with

$$t = \sqrt{A} \, . \, \frac{b-\beta_0}{s}, \tag{47}$$

where β_0 is any theoretical value of the coefficient, such as zero, proposed for comparison. Equally, the unknown β may be assigned a known frequency distribution in the light of the observations, based on

$$\beta = b + \frac{st}{\sqrt{A}}, \tag{48}$$

where t is distributed in "Student" 's distribution for $(N-2)$ degrees of freedom.

This simple and well-known form of analysis is not, I believe, disputed, save in the interpretation of the fiducial inferences. It should be noted none the less that it does violate the criterion of judging by repeated sampling, for in repeated sampling from the bivariate distribution of x and y, the value of A would vary from sample to sample. The distribution of $(b-\beta)$ would no longer be normal, and, before we knew what it was, the distribution of A, which in turn depends on that of x, would have to be investigated. Indeed, at an early stage Karl Pearson did attempt the problem of the precision of a regression coefficient in this way, assuming x to be normally distributed. The right way had, however, been demonstrated many years before by Gauss, and his method only lacked for completeness the refinement of the use of "Student" 's distribution, appropriate for samples of rather small numbers of observations.

To judge of the precision of a given value of b, by reference to a mixture of samples having different values of A, and therefore different precisions for the values of b they supply, is erroneous because these other samples throw no light on the precision of that value which we have observed. If we must think in terms of random sampling, it is only that *selection* of random samples which agree exactly with our own in respect to the value of

$$A=S(x-\bar{x})^2\,, \tag{49}$$

that is relevant to assessing its real precision. Such a selection might be quite inaccessible in sampling for acceptance. The function of selection is to transform the crude aggregate obtained by "repeated sampling" into a sample genuinely representative of the reference set appropriate to the test of significance. It is not obvious that this can always be done by selection from the material obtainable by mechanical repetition; when the test of significance is understood, such a roundabout course is unnecessary.

4. The two-by-two table

Although in the case of simple regression there has not, so far as I know, been any tendency to calculate erroneous values through mistaking the logical nature of the test, at least since the time of Karl Pearson, the very similar and equally fundamental test of proportionality in a two-by-two table has on more than one occasion become a matter of dispute. The data in this case consist of a number of cases doubly dichotomized, as people may be classified as male or female, or, again as tasters or non-tasters of phenylthiocarbamide, which a proportion of people cannot taste at concentrations which to others taste distinctly bitter. The statistical test is intended to find out whether the four frequencies observed are in proportion, or, in other words, whether the two classifications are independent. It will be noticed that, as in most tests, what is here to be rejected by a significant result is a whole class of hypotheses. These will have various values for the expected marginal frequencies, each, however, having proportional frequencies in the contents.

In this problem, as was pointed out concurrently by Yates[29] and Fisher[7] in 1934, a great simplification is available in that the subset of possible samples having the same marginal frequencies as that observed will have, whenever the two classifications are independent, the same frequency distribution. So, if in the records of a controlled experiment we find that three treated animals all die, and three control animals all survive, the two-by-two table

	Died	Survived	Total
Treated . .	3	0	3
Control . .	0	3	3
Total . .	3	3	6

is recognized to be one of a subset of possibilities, being double dichotomies having the same marginal totals, represented briefly by

	0 | 3		1 | 2		2 | 1		3 | 0	
	3 | 0		2 | 1		1 | 2		0 | 3	
Frequency	1		9		9		1	÷ 20

These possibilities have relative probabilities independent of the unknown frequencies of death or survival, and calculable merely by the algebraic method of permutations; so that within this subset the observed result is seen to be the most successful out of twenty possibilities, all equally probable on any view that the treatment is without effect.

In other cases of compound hypotheses it does not always occur that when all the possibilities to be excluded are excluded by a single test, the level of significance is the same as the frequency of erroneous exclusion, for all the possible hypotheses to be tested. It may then be that when some particular case of the compound hypothesis is true, the proportion of samples capable of dismissing the whole range of hypotheses under test will not be so great as the level of significance would suggest. It would indeed be unreasonable in general to expect it to be otherwise; it is therefore worth calling special attention to the exceptional feature of the subset available for testing significance in the two-by-two table, that the frequency of rejection is the same for all the simple hypotheses included.

On two occasions (Wilson, 1941,[26, 27, 28] Barnard, 1945[1], 1949[2]) distinguished mathematical statisticians have tried to improve the test by including in the enumeration cases in which the marginal totals differ from those in the sample observed. On both occasions, after discussion and elucidation of the logical basis of the test, these attempts have been abandoned. The argument was well illustrated by Barnard using the case discussed above, wherein, if we had to consider repetition of the experiment on the assumption that the treated and control animals have the same expectation p of dying, the frequency of what has been observed is exactly

$$p^3 q^3. \qquad (50)$$

For any real values of p and q, adding to unity, this is a small fraction. Its *maximum* value, attained when $p=\frac{1}{2}$, is only 1/64. So, if a repetition of the experiment were the right criterion, as has been very dogmatically asserted, significance at least at this level could be claimed. It was on this basis asserted that the method of Neyman and Pearson, relying on the formula of repeated sampling, had led to "a much more powerful test than Fisher's". The 64 cases enumerated in this argument include, however, not only the 20 having the same marginal totals, but 44 others belonging to six other subsets, of which it should be noted (i) that what has been observed does not belong to any of these six subsets, and (ii) within each of these questionable subsets there is no configuration which could be judged significant. Their inclusion in the enumeration can therefore do no other than enhance the apparent significance by inflation of the denominator.

Professor Barnard has since then frankly avowed that further reflection has led him to the same conclusion as Yates and Fisher, as indeed Wilson, with equal generosity, had done earlier. It is, therefore, obvious that he had at first been misled by the form of argument developed by Neyman and Pearson. E. S. Pearson himself has, so far as I know, made no such disavowal; but the numerical values given in Table 38 of his *Biometrika Tables* (1954)[21] do seem *in fact* to have been calculated by the method of Fisher and Yates, though no credit is given to these authors. It is to be feared, therefore, that the principles of Neyman and Pearson's "Theory of Testing Hypotheses" are liable to mislead those who follow them into much wasted effort and disappointment, and that its authors are not inclined to warn students of these dangers.

5. Excluding a composite hypothesis

It was remarked above that very commonly a test

of significance is used to exclude any one of a class of hypotheses, or, as it is sometimes called, a composite hypothesis. Some of the examples given have the exceptional feature that a single test can be found appropriate for the purpose, in which, were any of this class of hypothesis true, the criterion of rejection would be satisfied with the same frequency as the appropriate level of significance. The errors due to equating these concepts which have been indicated so far have been only of the kind which flow from ignoring the appropriate subset of cases to which the observed sample belongs, and seeking to ascertain the frequency of occurrences in a more inclusive set containing elements of a different kind, the variations of which are irrelevant to the observed case under consideration.

Composite hypotheses in general, however, contain another reason for ignoring the assumption that the frequency of rejection should be equated to the level of significance; which reason flows from the very fact that they are composite, i.e. that two or more distinct possibilities are to be rejected, each on sufficiently strong evidence. It may be that samples of the kinds available do not so easily dismiss the whole range of hypotheses to be tested even at a moderate level of significance.

A simple, though artificial, example is the following. A number of cards is made up into a pack, in which the proportions of the four suits are unknown. The composite hypothesis to be tested is that the proportions of the two red suits do not both exceed 25%. The data on which the test is to be based are a sample of 100 random draws each followed by replacement and reshuffling.

The possibility that there were no more than 25%

Hearts could be excluded at a reasonable level of significance, if 34 Hearts appeared in 100 chosen. If the true proportion were 25%, the probability of observing 34 or more is found to be 2·759%. However, the composite hypothesis is disproved only if it is demonstrable that the proportion of Diamonds also is more than 25%, and this requires that the sample should contain at least 34 Diamonds. It is not difficult to anticipate that both these conditions together will be fulfilled very rarely, even in the case in which both Hearts and Diamonds contribute a full quarter to the material sampled. In fact, an apparent disproof of the composite hypothesis at the moderate level of significance chosen would in this case be obtained in less than 34 trials in a million.

Even if the test had been based on 1000 randomly chosen cards, so that 278 were required of each red suit, a pack with exactly 25% of each would provide an apparent disproof with a frequency of only about 38 trials in a million.

It is, of course, no inconvenience that the frequency of rejecting the hypothesis in some cases when it is true should be low, but the calculation indicates also that even if both suits did really occupy somewhat more than 25% of the material sampled, it would not be easy to demonstrate this fact, even at a moderate level of significance, if neither of them were greatly in excess. Sufficiently large samples could indeed make such a demonstration probable; but the frequency of attaining a significant sample in the limiting case where both red suits have exactly 25%, is always less than the square of the fraction measuring the level of significance.

In scientific work it is necessary to be able to assess the strength of the evidence that a particular hypothesis, single or composite, appears to be untenable. The example has been chosen to show that strong evidence may sometimes be hard to obtain. Warnings that the strength of the evidence is not to be measured by the frequency observed in "repeated sampling from the same population" have not, on the whole, been well received by the authors of this formula. E. S. Pearson (*Biometrika*, 1947)[20] quotes me as writing (*Sankhya*, 1945)[14]:

> In recent times one often repeated exposition of the tests of significance, by J. Neyman, a writer not closely associated with the development of these tests, seems liable to lead mathematical readers astray, through laying down axiomatically, what is not agreed or generally true, that the level of significance must be equal to the frequency with which the hypothesis is rejected in repeated sampling of any fixed population allowed by hypothesis. This intrusive axiom, which is foreign to the reasoning on which the tests of significance were in fact based seems to be a real bar to progress. . . .

On this E. S. Pearson remarks, "But the subject of criticism seems to me less an intrusive mathematical axiom than a mathematical formulation of a practical requirement which statisticians of many schools of thought have deliberately advanced". It would, however, be difficult to find "schools of thought" other than that of Neyman and Pearson themselves, which have deliberately advanced anything of the kind. The rather wooden attitude adopted by this school seems to stem only from their having committed themselves to an unrealistic formalism.

Obvious as it might seem, it is evidently necessary to point out that it is no remedy to construct a test of significance with a firm intention that the hypothesis *shall* be rejected when true in a fixed proportion of trials. For this may well be mathematically impossible, for the whole range of cases; and consequently, what is of much greater importance, a test which is made to conform in one case may go widely astray in others. For example, if the test were chosen in the example discussed above, that the sum of the numbers of the two red suits observed in a sample of 100 should exceed 60 (which, if both red suits contributed just 25%, would reject the hypothesis to be tested with a "satisfactory" frequency, 2·623%), it would reject it very much more frequently in other cases, in which the null hypothesis was true, as for example if Hearts were 50% and Diamonds only 25%. Such a "test" while purporting to examine the truth of the null hypothesis would in fact reject it readily without the evidence of the observations being appreciably adverse. At least one of the Tables published by Professor E. Pearson and H. O. Hartley is indeed misleading in just this way (No. 11).[21] The editors' justification is: "It is the result of an attempt to produce a test which *does* satisfy the condition that the probability of the rejection of the hypothesis tested will be equal to a specified figure". The logical basis of a test of significance as a means of learning from experimental data is here completely overlooked; for the potential user has no warning of the concealed laxity of the test.[15]

In fact, as a matter of principle, the infrequency with which, in particular circumstances, decisive evidence is obtained, should not be confused with the force, or cogency, of such evidence. It is the highest frequency, within the range of the class of hypotheses tested, that is logically relevant.

6. Behrens' Test

The immediate effect of "Student" 's work, as its use came to be appreciated, was to supply a test of significance for the deviation, from some theoretically expected value, of the mean of a sample of observations, or of a regression, even though based on a small sample. It was quickly shown that the same Table could be used for the comparison of means, or regressions, based on small samples, provided the two sets of observations to be compared had the same precision, or to take a rarer but more general case, precisions in a known ratio.

These early results covered immediate practical requirements rather fully, for with large samples capable of supplying, on internal evidence, accurate estimates of precision, the large-sample procedure of estimating the two sources of error independently could be relied on; and, in a great deal of practical experimental work with small samples, although different lots of material might in reality have somewhat unequal variances, there were good reasons for supposing the real differences to be small compared with the errors of estimation from the small samples individually; so that better comparisons would be obtainable by pooling the variances of the different lots. The mathematical problem of the comparison of means of samples, not only small in size, but for which there is no reason *a priori* to dismiss the largest imaginable differences in precision, was of mathematical interest, and potential experimental importance, though it is not common to find realistic data which present this problem, partly because they are rarely sought.

For samples from a single population, the effect of eliminating the unknown variance, σ^2, by "Student" 's method, on the distribution of the error of the mean, is to replace, in the specification of this error,

$$\frac{\sigma}{\sqrt{N}} x , \tag{51}$$

where x is normally distributed with unit variance, but σ is unknown, by

$$\frac{s}{\sqrt{N}} t , \tag{52}$$

where t is distributed in "Student" 's distribution, for the appropriate number of degrees of freedom, $n(=N-1)$, and s is the estimate of σ available from n degrees of freedom.

For two samples from populations having a common mean, the deviations will be independent, and the data will supply values s_1, based on n_1 degrees of freedom, and s_2 based on n_2. The difference between the observed means is then the sum (or difference) of the two deviations from the true mean, so that on the null hypothesis considered, namely that the two population means are equal, we have

$$\bar{x}_1 - \bar{x}_2 = \frac{s_1}{\sqrt{N_1}} t_1 - \frac{s_2}{\sqrt{N_2}} t_2 , \tag{53}$$

where t_1 and t_2 are distributed independently in the two "Student" distributions.

If the frequency is small, such as 1%, that the expression on the right, which has a known distribution, for the observed values s_1 and s_2, shall exceed the observed difference in the sample means, this difference may be judged significant.

Effectively this solution, in a somewhat different

notation, was given by W.-U. Behrens[3] in 1929; Behrens also gave a short numerical table. A paper of mine in 1935 confirmed and somewhat extended Behren's theory. A much more substantial table was supplied by P. V. Sukhatmé[25] in 1938 covering values of n_1 and n_2 of 6, 8, 12, 24 and ∞, and values of the ratio $s_1/s_2(=\tan\theta)$ for values of θ of 0°, 15°, 30°, 45°, 60°, 75° and 90°. Apart from very small samples this was sufficient to make the test easily available. In 1941 the author gave[13] asymptotic expansions for calculating the probabilities with accuracy in any particular case, and a further range of tables for the case when either n_1 or n_2 is large. More recently Fisher and Healy[16] have given, for a number of levels of significance, the exact values when n_1 and n_2 are small and odd. The numerical values seem to make allowance nicely for the fact that a composite hypothesis, in which all ratios of σ_1/σ_2 are possible, is being tested, for it is required to set a limit which will rarely be passed by random samples of populations having the same mean, whatever may be the true variance ratio (*Statistical Tables* fifth edition VI, VI_1 and VI_2).[12]

In the extreme case in which both samples are of only two readings such as x_1 and x_2 for the first sample, and y_1 and y_2 for the second, Behrens' test takes the simple form of calculating

$$\frac{x_1+x_2-y_1-y_2}{|x_1-x_2|+|y_1-y_2|} = T, \tag{54}$$

where the level of significance is determined by giving T "Student" 's distribution for one degree of freedom, so that the level of significance is

$$\frac{2}{\pi}\tan^{-1}(1/T). \tag{55}$$

This extreme case, academic as it is, is particularly suitable for exhibiting the logic of the test. In 1937 I gave[11] the frequency distribution of T in repeated samples from populations having a fixed variance ratio,

$$\phi = \sigma_2^2/\sigma_1^2 \tag{56}$$

in the exact form,

$$dp = \frac{dT}{\pi(1+T^2)}\left\{\sqrt{\frac{\phi}{(1+\phi+T^2)}}+\sqrt{\frac{1}{1+\phi+\phi T^2}}\right\}, \tag{57}$$

with the probability integral,

$$\frac{1}{2}+\frac{1}{\pi}\left\{\sin^{-1}\frac{T\sqrt{\phi}}{\sqrt{(1+\phi)(1+T^2)}}+\sin^{-1}\frac{T}{\sqrt{(1+\phi)(1+T^2)}}\right\} \tag{58}$$

from which it is evident that, if we take the 5% value of $|T|$ as

$$\tan 85^\circ\, 30' = 12{\cdot}7062$$

this value will be exceeded in 5% of such repeated trials only at the limits $\phi=0$, or ∞, while midway between these, if ϕ were equal to unity, the criterion would be exceeded in less than 1% of random trials.

This circumstance, indeed, caused me no surprise, for the reference set in (58) has not been limited to the subset having the ratio s_1/s_2 observed, but was eagerly seized upon by M. S. Bartlett, as though it were a defect in the test of significance of a composite or comprehensive hypothesis, that in special cases the criterion of rejection is less frequently attained by chance than in others. On reflexion I do not think one should expect anything else, and it was perhaps only because at this time Bartlett was confidently putting forward an alternative attempt to solve the same problem that he made so much of a circumstance which is, indeed, generally to be expected.

7. The "randomization test"

How important it seemed to Bartlett that, whatever the true nature of the population sampled, the null hypothesis should be rejected when true with exactly the frequency suggested by the level of significance, is shown by the fact that he did for a time put forward as an alternative, presumably thought to be better than Behrens', a test involving a deliberately introduced element of hazard.

Observing the formal resemblance of the problem discussed by Behrens, in this extreme case when both samples are of two values only, to a paired comparison of "Student" 's type with only two pairs, it is easy to see that if in addition to Behrens' value T, we calculate also

$$\frac{x_1+x_2-y_1-y_2}{\left||x_1-x_2|-|y_1-y_2|\right|} = T' , \tag{59}$$

then, if one of the values T, T' is chosen by an equal chance, the one chosen will exceed an approrpiate criterion such as 12·7, in absolute value with a frequency equal to the level of significance adopted; and this for all possible variance ratios of the population sampled.

This proposal, which has perhaps now been abandoned (though at the time an equally faulty proposal was quickly put forward by Neyman) has two conspicuous objections, one of which is of general importance in that it applies to all "randomization tests" in the Natural Sciences. Namely, that if T and T' lie on opposite sides of the criterion, T' being always the larger, and a coin is thrown to decide which shall be chosen, it is then obvious at the time that the judgement of significance has been decided not by the evidence of the sample, but by the throw of the coin. It is not obvious how the research worker is to be made to forget this circumstance; and it is certain that he ought not to forget it, if he is concerned to assess the weight only of objective observational facts against the hypothesis in question. A real experimenter, in fact, so far from being willing to introduce an element of chance into the formation of his scientific conclusions, has been steadily exerting himself, in the planning of his experiments, and in their execution, to decrease or to eliminate by randomization all the causes of fortuitous variation which might obscure the evidence.

Consequently, whereas in the "Theory of Games" a deliberately randomized decision[8] (1934) may often be useful to give an unpredictable element to the strategy of play; and whereas planned randomization[10] (1935-1966) is widely recognized as essential in the selection and allocation of experimental material, it has no useful part to play in the formation of opinion, and consequently in the tests of significance designed to aid the formation of opinion in the Natural Sciences.

The second and specific objection to Bartlett's T', as a test of significance, is that it does not increase or decrease monotonically for changes in the weight of the evidence. For example, if y_1-y_2 is less than x_1-x_2, then an equal change in y_1 and y_2, taking them farther apart, will diminish the denominator of T', and actually increase its value; so indicating a higher level of significance, due to a *greater* discrepancy between two parallel observations.

In fact a practical worker who had calculated T, and T', could only regard them as providing evidence of significance, if both exceeded the minimum level, and since T' is never less than T, this implies simply the use of Behrens' test.

It is an indication of the remoteness from scientific application with which this problem must have been discussed that soon after Bartlett put forward his randomization test an equivalent test using pairing of the n_1 values of one sample, at random with an equal number from the other sample, as can be done in no less than

$$\frac{n_2!}{(n_2-n_1)!} \tag{60}$$

different ways, was put forward by J. Neyman as a general solution of the problem. That "solution" also has never, I believe, been applied in practice.

8. Qualitative differences

The examples elaborated in the foregoing sections of numerical discrepancies arising from the rigid formulation of a rule, which at first acquaintance it seemed natural to apply to all tests of significance, constitute only one aspect of the deep-seated difference in point of view which arises when Tests of Significance are reinterpreted on the analogy of Acceptance Decisions. It is indeed not only numerically erroneous conclusions, serious as these are, that are to be feared from an uncritical acceptance of this analogy.

An important difference is that Decisions are final, while the state of opinion derived from a test of significance is provisional, and capable, not only of confirmation, but of revision. An acceptance procedure is devised for a whole class of cases. No particular thought is given to each case as it arises, nor is the tester's capacity for learning exercised. A test of significance on the other hand is intended to aid the process of learning by observational experience. In what it has to teach each case is unique, though we may judge that our information needs supplementing by further observations of the same, or of a different kind. To regard the test as one of a series is artificial; the examples given have shown how far this unrealistic attitude is capable of deflecting attention from the vital matter of the weight of the evidence actually supplied by the observations on some theoretical question, to, what is really irrelevant, the frequency of events in an endless series of repeated trials which will never take place. The rejection of Behrens' test, and the series of futile attempts to find an alternative, based on the Neyman and Pearson theory of testing hypotheses, well exhibit the sterility of such formalism.

The concept that the scientific worker can regard himself as an inert item in a vast co-operative concern working according to accepted rules, is encouraged by directing attention away from his duty to form correct scientific conclusions, to summarize them and to communicate them to his scientific colleagues, and by stressing his supposed duty mechanically to make a succession of automatic "decisions", deriving spurious authority from the very incomplete mathematics of the Theory of Decision Functions. Even if this theory in its development so far had steered clear of confusion between the fields of tests of significance in the Natural Sciences, of policy or strategy in the Theory of Games, of rejection in Quality Control, and perhaps of other situations equally slightly related, it would still be true that the Natural Sciences can only be successfully conducted by responsible and independent thinkers applying their minds and their imaginations to the detailed interpretation of verifiable observations. The idea that this responsibility can be delegated to a giant computer programmed with Decision Functions belongs to the phantasy of circles rather remote from scientific research. The view has, however, really been advanced[19] (Neyman, 1938) that Inductive Reasoning does not exist, but only "Inductive Behaviour"!

A misconception having some troublesome consequences was introduced[18] by Neyman and Pearson in 1933, shortly after they had learnt of the possibility of deriving probability statements and therefore limits of significance by the fiducial argument, which had been published in the same journal, the *Proceedings of the Cambridge Philosophical Society*, in 1930.[5] Instead of perceiving that my method was appropriate to the absence of knowledge *a priori*, and, although I had not made this clear, would have been invalidated by the presence of such knowledge, Neyman and Pearson speak of my results as though they were a kind of "greatest common measure" of the inferences which could be made for all possible types of information *a priori*. In fact their paper opens as follows:

> In a recent paper[17] we have discussed certain general principles underlying the determination of the most efficient tests of statistical hypothesis, but the method of approach did not involve any detailed consideration of the question of *a priori* probability. We propose now to consider more fully the bearing of the earlier results on this question and in particular to discuss what statements of value to the statistician in reaching his final judgment can be made from an analysis of observed data, which would not be modified by any change in the probabilities *a priori*. In dealing with the problem of statistical estimation,* R. A. Fisher has shown how, under certain conditions, what may be described as

* My paper had, however, been entitled "*Inverse Probability.*"

rules of behaviour can be employed which will lead to results independent of these probabilities; in this connection he has discussed the important conception of what he terms fiducial limits.[5,6] But the testing of statistical hypotheses cannot be treated as a problem in estimation, and it is necessary to discuss afresh in what sense tests can be employed which are independent of *a priori* probability laws.

This early misconception has led other writers to seek for inferences independent of *a priori* laws, whereas seeing that Bayes' theorem is based upon supposedly exact knowledge of probabilities *a priori*, and that these probabilities can be made to appear explicitly in the result, none but trivial conclusions can be common to all cases. It is perhaps some sort of recognition of this that makes these authors ascribe to me "rules of behaviour", which I had not mentioned at all, whereas I had written in quite conventional terms which refer to reasoning processes, such as "learning by experience" and the "probability of causes". The logical distinction must in any case be stressed between possessing no information of a certain kind, and possessing such information, although it may be provisionally expressed in a generalized notation. The confusion of these situations is a serious trap, especially for mathematicians without experience in the Sciences.

It is important that the scientific worker introduces no cost functions for faulty decisions, as it is reasonable and often necessary to do with an Acceptance Procedure. To do so would imply that the purposes to which new knowledge was to be put were known and capable of evaluation. If, however, scientific findings are communicated for the enlightenment of other free minds, they may be put sooner or later to the service of a number of purposes, of which we can know nothing. The contribution to the Improvement of Natural Knowledge, which research may accomplish, is disseminated in the hope and faith that, as more becomes known, or more surely known, a great variety of purposes by a great variety of men, and groups of men, will be facilitated. No one, happily, is in a position to censor these in advance. As workers in Science we aim, in fact, at methods of inference which shall be equally convincing to all freely reasoning minds, entirely independently of any intentions that might be furthered by utilizing the knowledge inferred.

REFERENCES

1. G. A. Barnard (1945). A new test for 2×2 tables. *Nature*, vol. 156, pp. 177.
2. G. A. Barnard (1949). Statistical inference. *J. Roy. Stat. Soc.*, B, vol. 11, pp. 115-139.
3. W.-U. Behrens (1929). Ein Beitrag zur Fehlen-Berechnung bei wenigen Beobachtungen. *Landw. Jb.*, vol. 68, pp. 807-837.
4. R. A. Fisher (1924). On a distribution yielding the error functions of several well-known statistics. *Proc. Inter. Math. Cong.*, Toronto, vol. 2, pp. 805-813.
5. R. A. Fisher (1930). Inverse probability. *Proc. Camb. Phil. Soc.*, vol. 26, pp. 528-535.
6. R. A. Fisher (1933). The concepts of inverse probability and fiducial probability referring to unknown parameters. *Proc. Roy. Soc.*, A, vol. 139, pp. 343-348.
7. R. A. Fisher (1925-1958). *Statistical Methods for Research Workers.* Oliver and Boyd, Edinburgh.
8. R. A. Fisher (1934). Randomization, and an old enigma of card play. *Math. Gazette*, vol. 18, pp. 294-297.

9. R. A. Fisher (1935). The fiducial argument in statistical inference.
Ann. Eugen., vol. 6, pp. 391-398.

10. R. A. Fisher (1935-1966). *The Design of Experiments.*
Oliver and Boyd, Edinburgh. § 9 and 10.

11. R. A. Fisher (1937). On a point raised by M. S. Bartlett on fiducial probability.
Ann. Eugen., vol. 7, pp. 370-375.

12. R. A. Fisher and F. Yates (1938-1963) *Statistical Tables for Biological, Agricultural and Medical Research.*
Oliver and Boyd, Edinburgh.

13. R. A. Fisher (1941). The asymptotic approach to Behrens' integral, with further tables for the *d* test of significance.
Ann. Eugen., vol. 11, pp. 141-172.

14. R. A. Fisher (1945). The logical inversion of the notion of the random variable.
Sankhya, vol. 7, pp. 129-132.

15. R. A. Fisher (1956) On a test of significance in Pearson's *Biometrika Tables* (No. 11).
J. R. Stat. Soc. B., vol. 18, pp. 36-40.

16. R. A. Fisher and M. J. R. Healy (1956). New tables of Behren's test of significance.
J. R. Stat. Soc. B., vol. 18, pp. 212-216.

17. F. R. Helmert (1875). Ueber die Berechnung des wahrscheinlichen Fehlers aus einer endlichen Anzahl wahrer Beobachtungsfehler.
Z. Math. Phys., vol. 20, pp. 300-303.

18. J. Neyman and E. S. Pearson (1933). The testing of statistical hypotheses in relation to probabilities *a priori*.
Proc. Camb. Phil. Soc., vol. 29, pp. 492-510.

19. J. Neyman (1938). L'estimation statistique traité comme un problème classique de probabilité.
Actualités sci. industr., vol. 739, pp. 54-57.

20. E. S. Pearson (1947). The choice of statistical tests illustrated on the interpretation of data classed in a 2×2 table.
Biometrika, vol. 34, pp. 139-163.

21. E. S. Pearson and H. O. Hartley (1954). *Biometrika Tables for Statisticians.* Cambridge University Press.

22. K. Pearson (1900). On the criterion that a given system of deviations from the probable in the case of a correlated system of variables is such that it can be reasonably supposed to have arisen from random sampling.
Phil. Mag., Series V, vol. 1, pp. 157-175.

23. K. Pearson (1925). Further contributions to the theory of small samples.
Biometrika, vol. 17, pp. 176-199.

24. "Student" (1908). The probable error of a mean.
Biometrika, vol. 6, pp. 1-25.

25. P. V. Sukhatmé (1938). On Fisher and Behrens' test of significance for the difference in means of two normal samples.
Sankhya, vol. 4, pp. 39-48.

26. E. B. Wilson (1941). The controlled experiment and the four-fold table.
Science, vol. 93, pp. 557-560.

27. E. B. Wilson (1942). On contingency tables.
Proc. Nat. Acad. Sci., vol. 28, pp. 94-100.

28. E. B. Wilson and Jane Worcester (1942). Contingency tables.
Proc. Nat. Acad. Sci., vol. 28, pp. 378-384.

29. F. Yates (1934). Contingency tables involving small numbers and the χ^2 test.
Suppl. J. R. Stat. Soc. B., vol. 1, pp. 217-235.

CHAPTER V

SOME SIMPLE EXAMPLES OF INFERENCES INVOLVING PROBABILITY AND LIKELIHOOD

1. The logical consequences of uncertainty

The concepts sketched in Chapter III have arisen in the study of numerical observations in the Natural Sciences; they are intended for use in the inferences by which progress in the sciences is guided. Since the reasoning is quantitative it involves mathematical operations, which need not, however, be of a very complicated kind. Indeed, the examples may be confined to simple cases, though often at the expense of being scientifically trivial, for it is not the mathematics but the logical nature of these concepts that requires to be exemplified. Since the reasoning is inductive, the development for it of appropriate mathematical operations seems to run counter to the view that all mathematics can be reduced to a single and wholly deductive system. Admittedly deductive processes play a predominant part in mathematics, yet it is difficult to admit that mathematics is less than the whole art of exact quantitative reasoning, and as such must extend beyond the domain of deduction proper.

The theory that all mathematics could be reduced to a purely deductive system, which was popular about the beginning of this century, has, moreover, in the meantime suffered, with the development of axiomatic studies, some rather severe setbacks. It is common ground that the consistency of the axiomatic basis of a deductive system is essential for the reliability of its consequences. It has been formally demonstrated that a system admitting one contradiction must admit all, in the sense that any proposition whatever can be deduced from it, by formally rigorous processes. The non-existence of contradictory consequences is thus a burning question for the whole superstructure. Moreover, it has been proved that the non-existence of such contradictions can never be demonstrated on the basis of the axioms of the system themselves. It would be rather absurd, indeed, to imagine that any chain of theorems, derived from a given axiomatic basis, could disprove a possible property of that basis, when it is known that, if it had that property, these same theorems could certainly be deduced from it. For the possibility of proving such theorems does not depend upon the truth of what they assert. It would seem, therefore, that the validity of a purely deductive system has at best the same logical status as has a scientific theory, which has not yet been found in any case to be in conflict with the observations. As such it appears to be solidly based on a well-tested induction.

The axiomatic theory of mathematics has not been, and ought not to be, taken very seriously in those branches of the subject in which applications to real situations are in view. For, in applied mathematics, it is unavoidable that new concepts should from time to time be introduced as the cognate science develops, and any new definition having axiomatic implications is inevitably a threat to the internal consistency of the whole system of axioms into which it is to be incorporated. We have seen that the introduction of the concept of probability has caused just such an axiomatic disturbance which can only be remedied by a proper analysis of its meaning. In its applications, therefore, mathematics cannot easily be reduced to a closed and static system, but has to develop with the development of human thought, of which it is an important vehicle.

The purpose of deductive processes is to reveal, or uncover, the latent consequences of the axiomatic basis adopted. Nothing essentially new can be discovered, but the coherence of the whole structure can be usefully demonstrated, and its consistency to some extent tested. In the future these processes will perhaps be carried out better by machines than by men. The axiomatic basis, in any case, is tailored with a view to its deductive consequences, and it is this which gives it its real utility. Deductive arguments are, in fact, often only stages in an inductive process. For example, Bayes' theorem, on the data postulated, is strictly deductive, nevertheless we may include it among the processes of induction, on the ground that the probability statement *a priori* on which the argument is founded must in the real world have an inductive, and factual, rather than an axiomatic, basis.

On the contrary, the purpose of inductive reasoning, based on empirical observations, is to improve our understanding of the systems from which these observations are drawn. The appropriate mathematical forms for reasoning of this type have been becoming clear during the present century owing to the widespread application of statistical methods to scientific data, and of increasing understanding of the principles of the design of experiments. One of the obstacles which has had to be overcome is the tendency to impose on inductive thought the conventions and preconceptions appropriate only to deductive reasoning.

The governing characteristic of inductive reasoning is that it is always used to arrive at statements of uncertainty, and that logical situations are recognizable in which different types or degrees of uncertainty require to find rigorous expression. It has been thought that the Theory of Mathematical Probability, in spite of the fact that a probability statement is in reality a statement of a specific type of uncertainty, could be included among strictly deductive processes. This has seemed possible largely because many mathematical treatises have adopted a formal and abstract treatment in which the element of uncertainty is inoperative, just because applications to the real world are avoided.

The logical characteristic, which has been too much overlooked, of all inferences involving uncertainty is that the rigorous specification of the nature and extent of the uncertainty by which they are qualified must in general involve the whole of the data, quantitative and qualitative, on which they are based.

As soon as it is regarded realistically it is seen that the concept of Mathematical Probability shares this requirement. In a statement of probability the predicand, which may be conceived as an object, as an event, or as a proposition, is asserted to be one of a set of a number, however large, of like entities of which a known proportion, P, have some relevant characteristic, not possessed by the remainder. It is further asserted that no subset of the entire set, having a different proportion, can be recognized.

If, therefore, any portion of the data were to allow of the recognition of such a subset, to which the predicand belongs, a different probability would be asserted using the smallest such subset recognizable.

When no further subset is recognizable, which can be known only by an exhaustive scrutiny of the data, the predicand is spoken of as a random member of the ultimate set to which it belongs. An imagined process of sampling in which a succession of predicands are identified may be used to illustrate the relation between the proportion expected to be observed in the sample, and the primary proportion required to specify the set, now to be identified with the population sampled. Rather unsatisfactory attempts have been made to define the probability by reference to the supposed limit of such a random sampling process.

Difficulty has sometimes been expressed when the reference set, or the population sampled, is said to be infinite. The definition and consequent calculations can, however, be applied to any finite set however large, and the limit of these results, where the number in the set is increased indefinitely, is all that is meant by the results of sampling from an infinite population. The clarity of the subject has suffered from attempts to conceive of the "limit" of some physical process to be repeated indefinitely in time, instead of the ordinary mathematical limit of an expression of which some element is to be made increasingly great.

The following sections are intended to illustrate the kinds of reasoning, and concurrent mathematical operations, appropriate to various types of uncertainty.

2. Bayesian prediction

Expressed in terms of the hypothetical ratio $p : q$, Bayes' classical inference (from the uniform distribution *a priori* he used) is that the probability distribution of p is exactly

$$\frac{(a+b+1)!}{a!\,b!} p^a q^b \, dp \,, \tag{61}$$

in the light of the empirical observation of a successes out of $(a+b)$ trials. Alternatively, the hypothetical parameter p (or q) can be eliminated, and the inference expressed wholly in terms of the probability of future observations.

For example, if $c+d$ further trials were to be made with the same causal system, the probability, for each possible value of p, of observing just c successes is

$$\frac{(c+d)!}{c!\,d!} p^c q^d \,. \tag{62}$$

The average of this fraction, over all possible values of p, is then found by integration to be

$$\frac{(a+b+1)!}{a!\,b!} \cdot \frac{(a+c)!\,(b+d)!}{(a+b+c+d+1)!} \cdot \frac{(c+d)!}{c!\,d!} \,; \tag{63}$$

this represents the probability, in the light of the previous experience, of obtaining c successes in $(c+d)$ further trials, the hypothetical parameters p, q having been eliminated. It connects future rational anticipations directly with the experience on which they are based, without the mediation of hypothetical quantities. It is postulated only that the two samples are drawn from the same constant population of possibilities, and that Bayesian knowledge *a priori* is available.

It may be noticed that the last factor in the expression developed above,

$$\frac{(c+d)!}{c!\,d!} \tag{64}$$

stands only for the binomial coefficients forming the last line, or base, of Fermat's arithmetical triangle; but

$$\sum_{c=0}^{c+d} \frac{(c+d)!}{c!\,d!}\, p^c q^d \tag{65}$$

is not the only polynomial in p, q, the value of which is constantly equal to unity. If, in fact, the triangle is extended to any chosen boundary, as for example in the diagram (Fig. 3), the thirteen totals outside the boundary are the coefficients $\omega(c, d)$ of a polynomial

$$\Sigma\, \omega(c, d) p^c q^d = p^4 + 4p^6 q + 18p^6 q^2 + \cdots \tag{66}$$

of which the value is unity for all values of p.

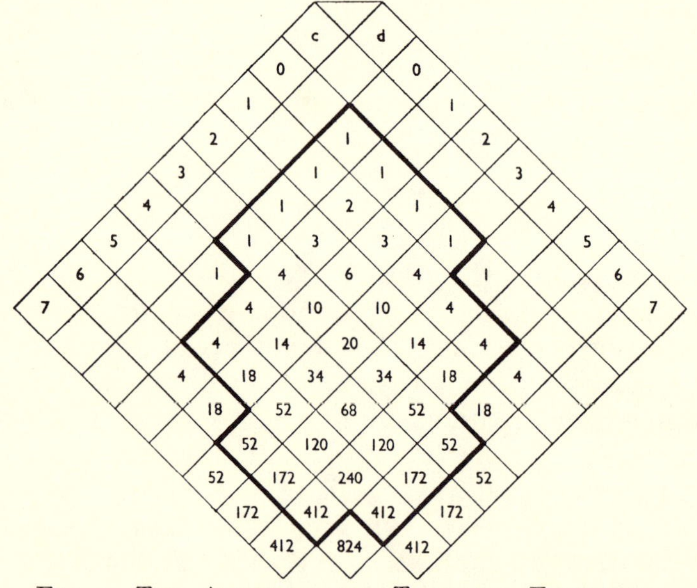

FIG. 3. THE ARITHMETICAL TRIANGLE EXTENDED

Then, based on previous experience of a successes out of $a+b$, we may infer the probability of reaching the terminal value (c, d) to be

$$\frac{(a+b+1)!}{a!\,b!} \cdot \frac{(a+c)!\,(b+d)!}{(a+b+c+d+1)!} \cdot \omega(c, d)\,, \qquad (67)$$

if a subsequent trial were made with these end-points.

3. Fiducial prediction

When there are no data *a priori* and the fiducial argument is available, the parametric values may equally be eliminated, and the appropriate inferences expressed as probability statements about future observations. In the case of the radioactive source considered in Section 3.3, if the total length of N_1 measured time intervals were X_1, we previously drew the inference that

$$\chi^2_{2N_1} = 2\theta X_1\,, \qquad (68)$$

where θ is the true rate of emission, and χ^2 is distributed as is the sum of $2N_1$ independent squares of normal deviates each having unit variance.

If, now, a second series of N_2 time readings were to give a total time of X_2 it follows equally that

$$\chi^2_{2N_2} = 2\theta X_2\,. \qquad (69)$$

The ratio of X_2 to X_1 is distributed, therefore, in random samples, in a distribution independent of θ, and depending only on N_1 and N_2. This distribution is the basis of the analysis of variance. The distribution of X_2 given X_1 is, in fact,

$$\frac{(N_1+N_2-1)!}{(N_1-1)!\,(N_2-1)!} \cdot \frac{X_1^{N_1} X_2^{N_2-1}\,dX_2}{(X_1+X_2)^{N_1+N_2}}\,, \qquad (70)$$

for all values of X_2 from 0 to ∞. Without discussing the possible values of the parameter θ, therefore, the exact probability of the total time recorded in a second series of trials lying within any assigned limits is thus calculable on the basis of the total time observed in the first series.

The probability of the value X_2 to be observed, exceeding any chosen value x, is the integral of the frequency element for all values of X_2 exceeding x, and is expressible as the sum of the first N_2 terms of a negative binomial expansion, i.e.

$$P = \left(\frac{X_1}{X_1+x}\right)^{N_1}\left\{1+\frac{N_1x}{X_1+x}+\frac{N_1(N_1+1)}{2}\left(\frac{x}{X_1+x}\right)^2+\right.$$
$$\left.\cdots+\frac{(N_1+N_2-1)!}{(N_1-1)!\,(N_2-1)!}\left(\frac{x}{X_1+x}\right)^{N_2-1}\right\}. \quad (71)$$

It should be observed that such fiducial probability statements about future observations are verifiable by subsequent observations to any degree of precision required. This is not possible for probability statements inferred about parametric values, save on the supposition that they are capable in some other way of direct observation. Probability statements about the hypothetical parameters are, however, generally simpler in form, and, once their equivalence is understood to predictions in the form of probability statements about future observations, they are seen not to incur any logical vagueness by reason of the subjects of them being relatively unobservable.

In carrying out such a verification as that suggested above, it is to be supposed that the investigator is not deflected from his purpose by the fact that new data are becoming available from which predictions, better than the one he is testing, could at any time be made. For verification, the original prediction must be held firmly in view. This, of course, is a somewhat unnatural attitude for a worker whose main preoccupation is to improve his ideas. It is perhaps for this reason that some teachers assert that statements of fiducial probability cannot be tested by observations. It is also to be noted that future events or repetitions of the same event, which would be independent for a fixed value of a parameter, will not generally be independent when the parameter has a frequency distribution.

4. Predictions from a Normal Sample

A case of particular interest of the fiducial probability of future observations is offered by the process of sampling a normal distribution. From a sample of N observations the two statistics, the estimated mean

$$\bar{x} = \frac{1}{N} S(x) , \tag{72}$$

and the estimated variance of the mean,

$$s^2 = \frac{1}{N(N-1)} S(x-\bar{x})^2 \tag{73}$$

subsume the whole of the information supplied by the sample about the population from which it was drawn. If μ is the true mean, the quantity,

$$\frac{\bar{x}-\mu}{s} = t , \tag{74}$$

has a distribution independent of the unknown parameters, well known to be

$$\frac{(\frac{1}{2}(N-2))!}{(\frac{1}{2}(N-3))!\sqrt{\pi(N-1)}} \cdot \frac{dt}{\left(1+\frac{t^2}{N-1}\right)^{\frac{1}{2}N}} . \tag{75}$$

for $N-1$ degrees of freedom.

This distribution has been very adequately tabulated so that the value of t is known for all levels of significance ordinarily required; the equation

$$\mu = \bar{x} - st \tag{76}$$

expresses μ as a random variable having a distribution of "Student" 's type, with $N-1$ degrees of freedom, and a scale factor s, calculable from the sample observed by a rigorous fiducial argument, provided no information *a priori* is available.

Ignoring, however, the mean, μ, of the hypothetical population, it would equally have been possible by a direct fiducial argument to calculate the distribution, in the light of the N observations already made, of a further observation, x, so far unknown. For if it is to be drawn at random from the same population, x will be normally distributed with variance N times that of $\bar{x}$, about the same mean, and independently of it, so that

$$x - \bar{x}$$

has a normal distribution about zero with variance $(N+1)$ times as great as has $\bar{x}$, and consequently when the observed value s is used to eliminate σ has the distribution specified by

$$x - \bar{x} = st\sqrt{N+1} \tag{77}$$

where, again, t has $N-1$ degrees of freedom. This prediction is evidently capable of verification to any degree of precision.

The same argument may be applied to predicting the values of a future sample of N', or, in particular of the statistics $\bar{x}'$ and s' derivable from it. In this case the value of the population variance could be estimated from the sum of the squares within both samples, that is as

$$\frac{1}{N+N'-2}\{S(x-\bar{x})^2+S'(x'-\bar{x}')^2\}, \tag{78}$$

and this quantity multiplied by

$$\frac{1}{N}+\frac{1}{N'} \tag{79}$$

would provide the estimated variance of the difference between the observed means

$$\bar{x}-\bar{x}'$$

based on $N+N'-2$ degrees of freedom. Consequently, for unselected samples

$$(\bar{x}-\bar{x}')\cdot\sqrt{\frac{NN'}{N+N'}}\cdot\sqrt{\frac{N+N'-2}{S+S'}} \tag{80}$$

will be distributed as is "Student" 's ratio t for $N+N'-2$ degrees of freedom. A second pivotal relationship is needed, since two values are to be predicted, and this is supplied by

$$2z=\log\frac{S}{N-1}-\log\frac{S'}{N'-1}, \tag{81}$$

the logarithm of the ratio of two independent estimates of the same variance, of which the distribution depends only on N and N', and is independent of the true mean.

To avoid misapplication of the method it should be noticed that for all pairs of primary observations $\bar{x}$, S, there is a one to one correspondence between pairs of "pivotal" values t, z, and corresponding pairs of predicted observations $\bar{x}'$, S'; equally for any pair of values $\bar{x}'$, S', a similar correspondence subsists between the pivotal values and the primary observations. Only on this condition the known frequency distribution of the pivotal values may be projected, or mapped, by direct substitution, to give the frequency distribution of the pair of unknowns to be predicted. The impossibility of statements of fiducial probability from discontinuous cases, such as the binomial distribution, is traceable to the fact that a single observational value corresponds, with any one parametric value, to a whole range in the values of the pivotal quantity, expressing the "probability integral" of the distribution.

In 1936,[4] addressing the Harvard tercentenary conference, I suggested that the condition for the further development of the use of fiducial inferences needed mathematical investigation, and would depend on the conditions of solubility of a type of problem, of which I gave, in general terms, an example, which has come to be known as the Problem of the Nile:

> The agricultural land of a pre-dynastic Egyptian village is of unequal fertility. Given the height to which the Nile will rise, the fertility of every portion of it is known with exactitude, but the height of the flood affects different parts of the territory unequally. It is required to divide the area, among the several households of the village, so that the yields of the lots assigned to each shall be in predetermined proportions, whatever may be the height to which the river rises.

The problem has not, I believe, in the meanwhile yielded up the conditions of its solubility, upon which, it would appear, the possibility of fiducial inference with two or more parameters, must depend.

It should, however, be noted, in the case of the two parameters of the normal distribution, that the sum of squares, the statistic S, not only yields a sufficient estimate for the true variance σ^2, but has a distribution independent of the true mean, μ, and so would supply also a solution of the type demanded in the Nile problem, if μ were the unknown variable.

An interesting application of the simultaneous distribution predicted for the two statistics of a future sample of N' values, is to allow N' to increase without limit, so that $\bar{x}'$ shall tend "in probability" to the population mean, μ, and the ratio

$$S'/(N'-1) \tag{82}$$

to the population variance, σ^2. We then have the simultaneous distribution, in the light of the first sample only, of the two parameters characterizing the population sampled. The frequency element of this simultaneous distribution is found to be the product

$$\sqrt{\frac{N}{2\pi\sigma^2}} \cdot e^{-\frac{N}{2\sigma^2}(\mu-\bar{x})^2} d\mu \, .$$

$$\frac{1}{\frac{N-3}{2}!} \cdot \left(\frac{S}{2\sigma^2}\right)^{\frac{1}{2}(N-1)} e^{-\frac{S}{2\sigma^2}} \cdot \frac{d\sigma^2}{\sigma^2} \, . \tag{83}$$

The rigorous step-by-step demonstration of the bivariate distribution by the fiducial argument would in fact consist first of the establishment of the second factor giving the distribution of σ given S, disregarding the other parameter, μ, and then of finding the first factor as the distribution of μ given $\bar{x}$ and σ. Several writers have adduced instances in which, when the formal requirements of the fiducial argument are ignored, the results of the projection of frequency elements using artificially constructed pivotal quantities may be inconsistent. When the fiducial argument itself is applicable, there can be no such inconsistency.

It will be noticed that in this simultaneous distribution (83) μ and σ^2 are not distributed independently. Integration with respect to either variable yields the unconditional distribution of the other, and these are naturally those obtainable by direct application of the fiducial argument, namely that

$$\frac{\mu-\bar{x}}{s} \tag{84}$$

is distributed as is t for $(N-1)$ degrees of freedom, while

$$\frac{S}{\sigma^2} \tag{85}$$

is distributed as is χ^2 for $(N-1)$ degrees of freedom. The distribution of any chosen function of μ and σ^2 can equally be obtained. The doubts expressed by Bartlett[1] on this point appear to be quite groundless.

It should, in general, be borne in mind that the population of parametric values, having the fiducial distribution inferred from any particular sample, does not, of course, concern any population of populations from which that sampled might have been in reality chosen at random, for the evidence available concerns one population only, and tells us nothing more of any parent population that might lie behind it. Being concerned with probability, not with history, the fiducial argument, when available, shows that the information provided by the sample about this one population is logically equivalent to the information, which we might alternatively have possessed, that it had been chosen at random from an aggregate specified by the fiducial probability distribution.

5. The fiducial distribution of functions of the parameters

A problem of this kind of some practical importance arises when it is desired to locate that value on a continuous scale which divides a hypothetical normal distribution in a given ratio. For example, to find the value which is only exceeded by one in forty of the population, of which we must judge by means of a randomly chosen sample of N measured individuals.

If μ and σ are the mean and standard deviation of the distribution, the point to be located may be represented by

$$\mu+\alpha\sigma \tag{86}$$

where, for the chosen frequency of one in forty, the value of α must be about 1·96, and is always known in terms of the frequency specified.

If $\bar{x}$ and s stand for the mean and standard deviation as estimated from the sample of observations, using for this purpose

$$Ns^2 = S(x-\bar{x})^2,$$

we may put

$$\bar{x}+as = \mu+\alpha\sigma\ , \tag{87}$$

and calculate the distribution of the quantity a in random samples, for, any sample must yield such a value. The distribution of a, which will depend on α, but not on the unknown parameters μ and σ, effectively exhibits in known terms the fiducial distribution of the particular linear function of the two parameters, chosen for examination. Equally, if some value a, were chosen with the intention of calculating from a succession of random samples, drawn perhaps from different populations, the value

$$\bar{x}+as\,, \tag{88}$$

then a knowledge of the sampling distribution of α for given a would display the distribution of the unknown deviate α, and therefore of the frequency ratio in which the true distribution had been partitioned.

The distribution of a for given α was first calculated for the Introduction to the *Mathematical Tables* (vol. 1)[2] of the British Association (1931) as an illustration of the appearance in statistical work of the function

$$I_n(x) = \frac{1}{\sqrt{2\pi}} \int_x^\infty \frac{(t-x)^n}{n!}\, e^{-\frac{1}{2}t^2}\, dt$$

$$= \frac{1}{\sqrt{2\pi}} \int_0^\infty \frac{t^n}{n!}\, e^{-\frac{1}{2}(t+x)^2}\, dt\,. \tag{89}$$

Using a sample of N observations as basis, it takes the form

$$\frac{(N-1)!}{2^{\frac{1}{2}(N-3)}\,.\,\frac{N-3}{2}!}\,.\,(1+a^2)^{-\frac{1}{2}N}\,.\,e^{-\frac{N\alpha^2}{2(1+a^2)}}\,.\,I_{N-1}\left(-\frac{a\alpha\sqrt{N}}{\sqrt{1+a^2}}\right) da\,, \tag{90}$$

while the distribution of α for given a is easily expressed by putting

$$\alpha = \frac{u}{\sqrt{N}} + \frac{a\chi}{\sqrt{N}} \tag{91}$$

where u is normally distributed with unit variance, and χ is distributed *independently of* u in the familiar distribution for $N-1$ degrees of freedom.

$$\frac{1}{\frac{N-3}{2}!}\,(\tfrac{1}{2}\chi^2)^{\frac{1}{2}(N-3)}\,.\,e^{-\frac{1}{2}\chi^2}\,.\,\chi\, d\chi\,. \tag{92}$$

The variable α has then the distribution of the sum of two variates distributed respectively in the normal distribution and in that of χ for $(N-1)$ degrees of freedom.

On a point of pure logic it may be noticed that the sampling distribution of a for given α, like that of α for given a, is entirely independent of the parameters of the distribution. In the fiducial distribution found for the parametric function

$$\mu+\alpha\sigma = \bar{x}+as$$

it is the introduction of the two sufficient statistics $\bar{x}$ and s, which brings in the requirement, that there must be no data *a priori* about the parameters, on which the fiducial argument relies. It requires that the observed statistics can be taken as random, and are unselected, and therefore representative values for the population from which they were obtained.

6. Observations of two kinds

It has been shown that observations of different kinds may justify conclusions involving uncertainty at different levels. It will be of some interest to consider the logical situation when observations of two such kinds are both available.

For example, let us suppose it to be possible to set a recorder to determine, for an exactly adjusted time interval, whether or no a charged particle has been received in that interval. It will not be supposed that the instrument will count them, if it receives more than one, but that it is capable of distinguishing the possible case of none, from the possible group of cases of "one or more."

Let p stand for the unknown probability of there being no particle, q the probability of there being one or more, then, if ξ is the time interval for which the instrument is set, and θ the unknown rate of delivery per unit of time, it appears that

$$p = e^{-\xi\theta}. \tag{93}$$

If then out of n trials it is observed that on a occasions no particle is recorded, while on b occasions there was one or more, we have a logical situation, without knowledge *a priori*, having a Likelihood Function,

$$e^L \propto p^a q^b, \tag{94}$$

in which expressions in θ can be substituted for p and q, to give the Mathematical Likelihood of any value of θ; but there is in the data no basis for making probability statements determining the probability that θ should lie between assigned limits.

Suppose, now, using the same supply of charged particles, it is possible to measure accurately a single randomly chosen time-interval between successive emissions. If this measured value is x_1, then for any given θ the distribution of x is

$$e^{-\theta x}\, \theta\, dx\,; \tag{95}$$

the probability of the random variable x exceeding the observed value, x_1, is

$$P = e^{-\theta x_1}, \tag{96}$$

and the fiducial distribution of θ is

$$dP = e^{-\theta x_1} x_1\, d\theta\,. \tag{97}$$

This fiducial distribution supplies information of exactly the same sort as that which Bayes showed how to obtain as a distribution of probability *a priori*. In fact

$$P = p^{x_1/\xi} = p^{\lambda}, \tag{98}$$

supplies the element of frequency

$$dP = \lambda p^{\lambda-1}\, dp\,, \tag{99}$$

needed to complete Bayes' method. The simultaneous probability of p lying in this range, and of giving rise to the frequencies observed, is then

$$\frac{(a+b)!}{a!\, b!} \cdot \lambda p^{a-1+\lambda}\, q^b\, dp\,; \tag{100}$$

the integral for all values of p over the range from 0 to 1 is

$$\frac{(a-1+\lambda)!\, b!}{(a+b+\lambda)!} \cdot \frac{(a+b)!}{a!\, b!} \lambda\,,^* \tag{101}$$

and the probability *a posteriori* is

$$\frac{(a+b+\lambda)!}{(a-1+\lambda)!\, b!} p^{a-1+\lambda}\, q^b\, dp\,. \tag{102}$$

If the observation x_1 were made first and the recorder set so that

$$\xi = x_1\,, \tag{103}$$

then λ would be unity, and we should have exactly Bayes' solution; but if it were thought better to set, for example,

$$\xi = 2x_1\,, \tag{104}$$

then λ would be $\frac{1}{2}$, and the probability distribution *a posteriori* would be

$$\frac{(a+b+\frac{1}{2})!}{(a-\frac{1}{2})!\, b!} p^{a-\frac{1}{2}}\, q^b\, dp\,. \tag{105}$$

* The factorial function, $x!$, has been generalized from positive integers only to all real numbers exceeding -1, by the Eulerian integral

$$x! = \int_0^\infty t^x e^{-t}\, dt\,.$$

If the experimenter were bound to use a single setting, so that, after choosing ξ, the same time interval will be used for all tests with the same source, then the second choice, $\lambda=1/2$, will be on the average more informative (·4144 against ·4041) than the first proposal, $\lambda=1$. Between them there is an optimum with ξ about 51% above x; this makes the amount of information expected per trial about 2% higher still (·4232).

When the value of λ is open to choice, the question may arise as to what value it will be best to choose. There is here a problem in Experimental Design.

Supposing that the same value of ξ, and therefore of $\lambda(=x_1/\xi)$, is to be used in all subsequent trials, let

$$p = e^{-m},$$

so that

$$m = \xi\theta,$$

or

$$\log m = \log \xi + \log \theta,$$

and the amount of information supplied by each subsequent trial about $\log \theta$ will be equal to that supplied about $\log m$, since $\log \xi$ is a known constant. But the amount of information about $\log m$ given by each trial is known to be
(*Design of Experiments*, Section 68)

$$\frac{m^2}{e^m - 1} = i_{\log m}$$

The average value of this amount of information can be determined in terms of λ, and the value of λ which maximizes this average is the most profitable on the supposition adopted.

Since

$$P = p^\lambda = e^{-\lambda m}$$

the element of frequency is

$$\lambda e^{-\lambda m} dm,$$

and the function of λ to be maximized is

$$\lambda \int_0^\infty \frac{m^2 e^{-m\lambda}}{e^m - 1} dm$$

$$= \lambda \sum_{n=1}^{\infty} \int_0^\infty m^2 e^{-m(\lambda+n)} dm$$

$$= 2\lambda \sum_{n=1}^{\infty} \frac{1}{(n+\lambda)^3}$$

or

$$2\lambda\, s_3(\lambda),$$

where s_3 stands for the sum of the infinite series

$$\frac{1}{(1+\lambda)^3} + \frac{1}{(2+\lambda)^3} + \dots$$

The equation for the value of λ which maximizes the amount of information expected is therefore

$$s_3 - 3\lambda s_4 = 0$$

where s_4 stands for the sum of the series

$$\frac{1}{(1+\lambda)^4} + \frac{1}{(2+\lambda)^4} + \dots$$

I am indebted to Mr. R. H. Simpson of the Department of Statistics at Rothamsted for a rather exact solution of the equation, namely

$$\hat{\lambda} = \cdot 66142593,$$

or

$$\frac{1}{\lambda} = \frac{\xi}{x_1} = 1\cdot 5118851,$$

at which the amount of information about log θ added by each further trial is nearly

$$\hat{i} = \cdot 41440,$$

which may be compared with the absolute maximum

$$\hat{i} = \cdot 64761,$$

obtainable with the most favourable value of p, namely

$$p = \cdot 203188,$$

as has been shown in Section 68 of *The Design of Experiments*.

Such a distribution *a posteriori*, whether expressed in terms of p or of θ, would be logically equivalent to a Bayesian probability *a posteriori*, or, equally, to one based exclusively on a fiducial argument, for the parameter is in each case a random variable of known distribution.

The percentile values of p in this distribution correspond each to each with the percentile values of the fiducial distribution of θ, through the relation

$$\theta = \frac{1}{\xi} \log 1/p \,.$$

The percentile values *a posteriori*, like those of the partial sum of the binomial expansion, may be found from the Table of z (Table V) of *Statistical Tables*, for the particular percentiles there tabulated, and their complements. The values of degrees of freedom (n_1, n_2) are equated to $2b+2$ and $2a+2\lambda$, so that if 2λ is not an integer, the latter would require interpolation; for this the Table of z is better suited than the associated Table of the Variance Ratio, e^{2z}.

In mathematical teaching the mistake is often made of overlooking the fact that Bayes obtained his probabilities *a priori* by an appropriate *experiment*, and that he specifically rejected the alternative of introducing them axiomatically on the ground that this "might not perhaps be looked at by all as reasonable"; moreover, he did not wish "to take into his mathematical reasoning anything that might admit dispute".

In many continental countries this distinction, which Bayes made perfectly clear, has been overlooked, and the axiomatic approach which he rejected has been actually taught as Bayes' method. The example of this Section exhibits Bayes' own method, replacing the billiard table by a radioactive source, as an apparatus more suitable for the 20th century.

7. Inferences from likelihoods

The mode of inference which takes the form of probability statements about parameters can lead to the alternative mode in the form of probability statements about future verifiable observations, by a general form of calculation; namely, if

$$F(\theta)\,d\theta$$

is the probability that the parameter lies in the range $d\theta$, and if for values within this range the probability of any future observable contingency, A, is

$$p_A(\theta)$$

then, eliminating θ, the probability of A is

$$P_A = \int p_A(\theta)\,F(\theta)\,d\theta\,, \qquad (106)$$

taken over all possible values of the parameter.

The converse process of inferring the frequency distribution of θ from a knowledge of quantities such as P_A, for a sufficient variety of future possibilities A, is usually possible, and is extremely simple when A is taken to represent experience so ample as to confine the uncertainty of the parameter consistent with it, to an arbitrarily small range.

The cases in which the observations, together with other data, allow only of a statement of the Mathematical Likelihood, require separate consideration, to obtain a clear view of the rational prospect which future contingencies present, when knowledge of the Likelihood Function only is available.

In the previous Section it has been shown that, where data of two kinds are simultaneously available, one capable of supplying Likelihood statements only, while the other, however meagre and uninformative it may be quantitatively, is capable of leading to Probability statements, then the two kinds of data, available, as it were, in parallel, may, if exhaustive estimation is possible, supply inferences in terms of Mathematical Probability. In the case, also, in which the probability statement is based on an independent act of random sampling, this can be done, as Bayes had shown, by multiplying each element of the frequency distribution by a multiple of the corresponding Likelihood, this multiple being chosen to make the elements so formed add, or integrate, to unity.

Further, from the nature of the Likelihood Function it is evident that if data yielding Likelihood statements only are available from two independent sources, the aggregate of the two sources of data will supply simply a Likelihood Function found by multiplying together the two functions supplied by its parts.

In view of these various relationships it would be natural to expect that when the two types of nexus represented by Mathematical Likelihood and Mathematical Probability are connected, as it were, in series, so that we have the likelihood of an exhaustive set of possibilities, for each of which the probability of some event A, is known, the whole would yield statements, not of probability, but of likelihood only. This humbler status is not, however, incompatible with substantial utility.

Let us consider, from this standpoint, data of the Bayesian or Bernouillian type, without knowledge *a priori*, in which a successes have been observed out of $(a + b)$ trials, in relation to the rational bearing of such an observation on the prospect of observing c successes out of a subsequent $(c + d)$ trials. For example, if 3 successes have been observed out of 19 trials, what is the prospect of observing 14 successes in a subsequent set of 21 trials?

Considering the two-by-two table

3	16	19
14	7	21
17	23	40

we may recall that the relative likelihood of the first observation of three successes to sixteen failures is

$$p^3q^{16} \cdot \frac{19^{19}}{3^3 \cdot 16^{16}}, \qquad (107)$$

which can be raised to unity by an appropriate choice of the ratio $p : q$. Similarly, with the second sample, if the probabilities of success and failure are p' and q', the relative likelihood is

$$p'^{14}q'^{7} \cdot \frac{21^{21}}{14^{14} \cdot 7^7}; \qquad (108)$$

now, if p and p' are to be the same, so that the two samples are drawn fairly from the same population, the most likely common value to give them is the total ratio of success, or 17/40. Inserting this, and the complementary value, for p, q and for p', q', we find the likelihood of the whole table, namely the expression

$$\frac{17^{17} \cdot 23^{23}}{40^{40}} \cdot \frac{19^{19} \cdot 21^{21}}{3^3 \cdot 7^7 \cdot 14^{14} \cdot 16^{16}} \qquad (109)$$

in which the function

$$x^x \qquad (110)$$

has replaced the factorial function ($x!$) used in the expression for the probability of the numerical table discussed, among others having the same margins. Taking logarithms to the base 10, the numerical values are

x	$\log x^x$	x	$\log x^x$
17	20·9176317	3	1·4313638
19	24·2963184	7	5·9156863
21	27·7666052	14	16·0457925
23	31·3197402	16	19·2659197
	104·3002955	40	64·0823997
	106·7411620		106·7411620 .
	997·5591335		

The relative likelihood of the two-by-two table under discussion is only about ·003623, and this is the likelihood assignable to the prospective contingency of fourteen successes out of twenty-one, in view of the data available. With the aid of a Table of x^x, such as is given on page 139, it is easy to calculate such a series as the following:

Number of successes out of 21	Likelihood
14	·0036
13	·0093
12	·0216
11	·0460
10	·0904
9	·1642 .

Above 10 the likelihood is less than 1/15, and such future contingencies may be recognized in advance as definitely unlikely; above 8 the likelihood is still less than 1/5. The two values 9 and 10 thus lie in a zone in which the likelihood is still low. The test of significance discussed in Chapter IV, Section 4, could be used to confirm these judgements from another standpoint. The test of significance suffers, like the "confidence limits" calculated for a binomial distribution from some insensitivity due to the discontinuity of the distribution, and this may be thought to outweigh the rather formal advantage of asserting significance at a given level, when this yields no probability statement more definite than an inequality.

The likelihood assigned to a fourfold table is symmetrically related to the first (real) and second (conjectural) sample. The likelihood of observing 14 successes out of 21 as judged by data showing 3 successes out of 19, is exactly the same as the likelihood in prospect of observing 3 successes out of 19, judged on the basis of experience of 14 successes out of 21. Under another aspect, supposing both samples to have been observed, the same measure may be taken to be the likelihood of the hypothesis that they have been drawn from equivalent populations.

As in the cases in which probability rather than likelihood can be predicated, we may recover the likelihood statement appropriate to the parameters, by considering the likelihood in the limit of a very large sample showing in all pN successes out of N. The likelihood could then be written

$$\frac{(pN)^{pN}(qN)^{qN}}{N^N} \cdot \frac{19^{19}(N-19)^{N-19}}{3^3 16^{16}(pN-3)^{pN-3}(qN-16)^{qN-16}} \qquad (111)$$

leading, when N is increased without limit to

$$\frac{19^{19}}{3^3 \,.\, 16^{16}} p^3 q^{16} \qquad (112)$$

the likelihood as inferred directly for the parameters.

It may be noted that the likelihood of a future trial yielding c successes and d failures does not involve the factor

$$\frac{(c+d)!}{c!\,d!}, \qquad (113)$$

representing the number of ways in which such an outcome could occur. Nor, if a subsequent trial were to be made with chosen end-points, as in Fig. 3. would the likelihoods of these end-points involve the coefficients representing the number of paths in the extended triangle leading to each. Unlike a probability, the likelihood is independent of the number of ways in which the result could be brought about, for a statement of Likelihood does not involve a measureable reference set.

8. Variety of logical types

It is a noteworthy peculiarity of inductive inference that comparatively slight differences in the mathematical specification of a problem may have logically important effects on the inferences possible. In complicated cases, such effects may be very puzzling, as the conditions of solubility of problems of the Nile type have proved themselves to be. It may therefore be useful to consider some cases of extreme simplicity.

Let us suppose that x and y are two observed quantities, known each to be normally, and independently distributed with unit variance, x about an unknown value ξ, and y about η. It will be required to draw inferences about the pair of values (ξ, η) which may, of course, be represented as an unknown point, H, on a plane, on which (x, y) may be represented by an observed point, O.

If the data were as above, without further restriction, the probability distribution of the unknown point, formally demonstrable by the fiducial argument, is evidently a normal bivariate distribution, with unit variance in all directions, centred at the observed point (x, y). Additional data do, however, alter the character of the problem. It may be interesting to compare three cases:

(*a*) H is known to lie on a given straight line.

(*b*) H is known to lie on a circle.

(*c*) The given functional relationship between ξ and η does not confine this point either to a straight line, or to a circle, but to some other plane curve.

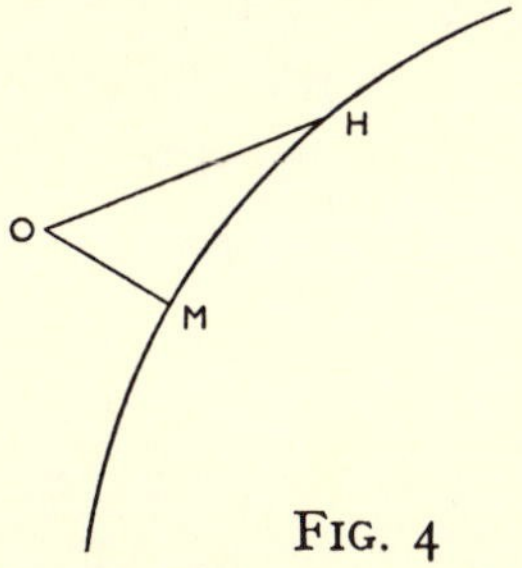

Fig. 4

In all cases the likelihood of any pair of parametric values is

$$e^{-\frac{1}{2}r^2},$$

where r is the distance from O to H. The nearest point on the curve to the observed point O then represents the solution of maximum likelihood:

If M stands for this point, the relative likelihood of any other point is

$$exp\{-\tfrac{1}{2}(OH^2-OM^2)\}.$$

Setting this equal to any chosen series of conventional fractions we shall have defined zones around the point of maximal likelihood M, comprising all points satisfying the functional relationship, however these may be connected.

(*a*) In the particular case in which H is restricted to a given straight line, M satisfies the condition for a Sufficient estimate, for the relative likelihood of any point H on the line is simply

$$exp(-\tfrac{1}{2}HM^2) \qquad (114)$$

and this is the same for all possible observation points on the line OM, produced if necessary. That is, for all observations leading to the same estimate.

The probability that HM should exceed any quantity u, positive or negative is therefore

$$\frac{1}{\sqrt{2\pi}}\int_u^\infty e^{-\frac{1}{2}t^2}\,dt\,; \qquad (115)$$

otherwise stated, the fiducial distribution of H is Normal with unit variance, and centred at the estimate point M.

In view of the relation (115) it appears that if the data were modified so that instead of an unlimited straight line, the possible range for H had a terminus T, then necessarily the fiducial distribution extends only so far as T, and at T it has a probability condensation equal to

$$\frac{1}{\sqrt{2\pi}} \int_{TM}^{\infty} e^{-\frac{1}{2}t^2}\, dt\,, \tag{116}$$

where TM is positive if M falls within the permitted range, but is taken to be negative if M lies outside it.

Knowing the frequency distribution of H, having coordinates (ξ, η), it is possible to calculate that of a second pair of random values (x', y'), the coordinates of a point O'. In the case of an unlimited straight line, it is easy to see that O' is distributed normally about M as centre, with variance unity in directions at right angle to the line, and twice as much in directions parallel with it. The contours of equal frequency density are ellipses with eccentricity $1/\sqrt{2}$, or ·707. The probability, on the data, of such a second trial lying within any defined area is then calculable.

(*b*)

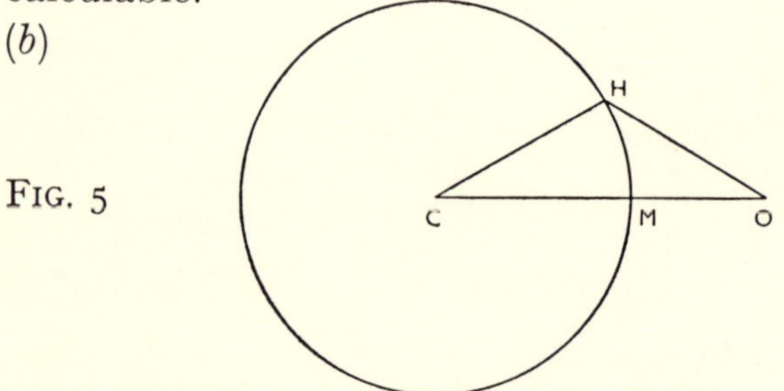

FIG. 5

The restriction which confines the unknown point H to the circumference of a given circle does not allow of a Sufficient estimate, for the relative likelihoods of the points on the circumference are not the same for different points O lying on the same radius CM, produced if necessary.

It is, however, a case in which an ancillary statistic can be recognized, namely the distance OC between the centre of the circle, and the point observed. For the sampling distribution of this distance must be the same for all points H on the circle. Consequently, we have only to determine the frequency distribution of the angle HCO, for given values of the distance OC, in order to obtain a fiducially determined distribution of the unknown point H.

If R stand for the radius of the circle, θ for the angle HCO, and a for the distance OC, then, since

$$HO^2 = R^2 + a^2 - 2aR\cos\theta\ , \tag{117}$$

the factor of

$$e^{-\frac{1}{2}HO^2} \tag{118}$$

which depends on θ is simply

$$e^{aR\cos\theta}\ . \tag{119}$$

But

$$\int_0^{2\pi} e^{aR\cos\theta}\,d\theta$$

$$= 2\pi\left(1 + \frac{a^2R^2}{2^2} + \frac{a^4R^4}{2^2\,.\,4^2} + \qquad\right) \tag{120}$$

$$= 2\pi I_0(aR)$$

expressed as a Bessel function in Bassett's notation. Hence, the fiducial distribution of θ is

$$\frac{d\theta}{2\pi I_0(aR)}\,e^{aR\cos\theta}\ , \tag{121}$$

the frequency density decreasing exponentially in the direction parallel to OC, at a rate, however, which depends on the ancillary statistic a. Here again we have a well-determined frequency distribution for the unknown point H (ξ, η), from which fiducial limits at all levels of probability can be calculated.

(*c*) In general, however, it is not to be expected either that a Sufficient statistic should exist, or that the most likely estimate could be made exhaustive by means of ancillary values. In such cases rational inference is effectively completed by the calculation of the Mathematical Likelihood for each plausible position of the unknown point. The practice of ignoring this quantity as a measure of rational belief appropriate to such cases would seem to leave the statistician who chooses this course without valid resource in respect of a great many subjects in which rational inference is possible. The fact that stronger inferences may be desired, and are certainly possible in other cases, seems to be no reason for not attuning our minds towards understanding the information actually available.

It is particularly to be noted in this example that the differences in the logical form of the available inferences flow from quite simple differences in the mathematical specification of the problem. Not to formal principles only, but there is also needed attention to particular analytic details.

REFERENCES

1. M. S. Bartlett (1937). Properties of sufficiency and statistical tests. *Proc. Roy. Soc.*, A, vol. 160, pp. 268-282.
2. R. A. Fisher (1931). The sampling error of estimated deviates. *Math. Tables*, vol. 1, pp. xxvi-xxxv.
3. R. A. Fisher (1935). The fiducial argument in statistical inferences. *Ann. Eugen.*, vol. 6, pp. 391-398.
4. R. A. Fisher (1936). Uncertain inference. *Proc. Amer. Acad. Arts and Sci.*, vol. 71, pp. 245-258.

	$x \log_{10} x$	$x^x 10^{-c}$		$x \log_{10} x$	$x^x 10^{-c}$
1	0	1	41	66·124 1381	1·3309
2	0·602 0600	4	42	68·176 4702	1·5013
3	1·431 3638	2·7	43	70·239 1436	1·7344
4	2·408 2400	2·56	44	72·311 9178	2·0508
5	3·494 8500	3·125	45	74·394 5631	2·4806
6	4·668 9075	4·6656	46	76·486 8603	3·0680
7	5·915 6863	8·2354	47	78·588 5993	3·8779
8	7·224 7199	1·6777	48	80·699 5794	5·0070
9	8·588 1826	3·8742	49	82·819 6079	6.6010
10	10·000 0000	1·0000	50	84·948 5002	8·8818
11	11·455 3195	2·8531	51	87·083 0190	1·2107
12	12·950 1750	8·9161	52	89·232 1739	1·7068
13	14·481 2636	3·0288	53	91·386 6211	2·4357
14	16·045 7925	1·1112	54	93·549 2630	3·5421
15	17·641 3689	4·3789	55	95·719 9479	5·2474
16	19·265 9197	1·8447	56	97·898 5295	7·9164
17	20·917 6317	8·2724	57	100·084 8668	1·2158
18	22·594 9051	3·9346	58	102·278 8236	1·9003
19	24·296 3184	1·9784	59	104·480 2687	3·0218
20	26·020 5999	1·0486	60	106·689 0750	4·8874
21	27·766 0652	5·8426	61	108·905 1199	8·0375
22	29·533 2990	3·4143	62	111·128 2847	1·3436
23	31·319 7402	2·0880	63	113·358 4546	2·2827
24	33·125 0698	1·3337	64	115·595 5183	3·9402
25	34·948 5002	8·8818	65	117·839 3682	6·9083
26	36·789 3070	6·1561	66	120·089 8997	1·2272
27	38·646 8216	4·4343	67	122·347 0118	2·2234
28	40·520 4249	3·3146	68	124·610 6061	4·0795
29	42·409 5419	2·5677	69	126·880 5873	7·5960
30	44·313 6376	2·0589	70	129·156 8628	1·4350
31	46·232 2125	1·7069	71	131·439 3428	2·7501
32	48·164 7993	1·4615	72	133·727 9397	5·3449
33	50·110 9600	1·2911	73	136·022 5688	1·0533
34	52·070 2832	1·1757	74	138·323 1473	2·1045
35	54·042 3816	1·1025	75	140·629 5948	4·2618
36	56·026 8900	1·0639	76	142·941 8330	8·7465
37	58·023 4638	1·0555	77	145·259 7858	1·8188
38	60·031 7767	1·0759	78	147·583 3790	3·8760
39	62·051 5197	1·1260	79	149·912 5402	8·1760
40	64·082 3997	1·2089	80	152·247 1990	1·7668

	$x \log_{10} x$	$x^x 10^{-c}$		$x \log_{10} x$	$x^x 10^{-c}$
81	154·587 2865	3·8662	116	239·477 1267	3·0000
82	156·932 7359	8·5652	117	241·977 7458	9·5005
83	159·283 4817	1·9208	118	244·482 0769	3·0344
84	161·639 4600	4·3597	119	246·990 0884	9·7744
85	164·000 6087	1·0014	120	249·501 7496	3·1750
86	166·366 8668	2·3274	121	252·017 0298	1.0400
87	168·738 1750	5·4724	122	254·535 8993	3·4348
88	171·114 4752	1·3016	123	257·058 3287	1·1437
89	173·495 7106	3·1312	124	259·584 2890	3·8396
90	175·881 8258	7·6177	125	262·113 7516	1·2994
91	178·272 7667	1·8740	126	264·646 6887	4·4329
92	180·668 4801	4·1541	127	267·183 0726	1·5243
93	183·068 9142	1·1720	128	269·722 8761	5·2951
94	185·474 0182	2·9786	129	272·266 0726	1·8453
95	187·883 7425	7·6514	130	274·812 6358	6·4958
96	190·298 0384	1·9863	131	277·362 5397	2·3043
97	192·716 8582	5·2102	132	279·915 7589	8·2368
98	195·140 1554	1·3809	133	282·472 2683	2·9667
99	197·567 8843	3·6973	134	285·032 0430	1·0766
100	200·000 0000	1·0000	135	287·595 0587	3·9360
101	202·436 4588	2·7955	136	290·161 2915	1·4497
102	204·877 2175	7·5373	137	292·730 7177	5·3792
103	207·322 2341	2·1001	138	295·303 3139	2·0059
104	209·771 4673	5·9084	139	297·879 0572	7·5693
105	212·224 8764	1·6783	140	300·457 9250	2·8703
106	214·682 4217	4·8131	141	303·039 8949	1·0962
107	217·144 0642	1·3934	142	305·624 9449	4·2164
108	219·609 7656	4·0716	143	308·213 0534	1·6333
109	222·079 4883	1·2008	144	310·804 1989	6·3709
110	224·553 1954	3·5743	145	313·398 3603	2·5024
111	227·030 8506	1·0736	146	315·995 5169	9·8973
112	229·512 4185	3·2540	147	318·595 6482	3·9414
113	231·997 8641	9·9509	148	321·198 7339	1·5803
114	234·487 1530	3·0701	149	323·804 7540	6·3790
115	236·980 2516	9·5555	150	326·413 6889	2·5923

CHAPTER VI

THE PRINCIPLES OF ESTIMATION

1. Relations to other work

The logical principles of statistical reasoning, which it is my purpose in this book to set out for explicit consideration, have underlain and been implicitly required in the development of the two other main aspects of Statistical Science, namely (*a*) the mathematical methodology of the handling of bodies of observational data, so as to elicit what they have to tell us, and (*b*) the Design of the logical structure of an observational record, whether of an experiment, or of a survey, so as to ensure its completeness and cogency as a tool of research. In the two books that I have written with these ends in view it has not in either case seemed appropriate to enlarge upon purely logical considerations which had in fact found their fullest expression in earlier work on the Theory of Estimation. This theory is adverted to, therefore, in these books only with their particular ends in view. In *Statistical Methods* to exhibit the existence of competent and practical methods applicable to data of many types, to exemplify some of the kinds of complication which ordinarily arise, and to bring a wider class of cases into logical connection with the Analysis of Variance. In the *Design of Experiments*, I had chiefly in view, in this part of the book, the use of the concept of Amount of Information as a measurable characteristic by which the precision of an experiment could be anticipated, or confirmed, and compared with the expenditure of effort entailed.

In both books I hoped that the examples exhibited would not only indicate methods useful in themselves, but would also facilitate the development of principles of reasoning by which a body of data can be interpreted. And I believe they have indeed had this effect. Nevertheless, in both cases, it was my object to set out only what was immediately serviceable to a particular end, and to avoid the multiplicity of abstract concepts which an adequate discussion of the subject from a logical standpoint inevitably requires. I shall hope in this Chapter, on the contrary, to direct attention primarily to the logical aspects, and to develop these with a minimum of mathematical and technical complexity.

The Theory of Estimation discusses the principles upon which observational data may be used to estimate, or to throw light upon the values of theoretical quantities, not known numerically, which enter into our specification of the causal system operating. These principles have been more or less familiar for many years, but have been confused by a number of false starts due to insufficient appreciation of the nature of the problem.

A primary, and really very obvious, consideration is that if an unknown parameter θ is being estimated, any one-valued function of θ is necessarily being estimated by the same operation. The criteria used in the theory must, for this reason, be invariant for all such functional transformations of the parameters. This consideration would have eliminated such criteria as that the estimate should be "unbiased", meaning that the average value of the estimate should be equal to the true estimand; for if this were true of any parameter, it could not also be true of, for example, its square.

Although "unbiased" in this small-sample sense is thus a comparatively useless concept, a definition serviceable asymptotically in large-sample theory may be given in the form that

$$Pr\{N^{\alpha}|T - \theta| > \varepsilon\}$$

shall tend to zero as N is increased without limit, whenever $\alpha \leqslant \frac{1}{2}$, where ε is a positive quantity, however small, and T is the proposed estimate of the parameter θ. Another criterion in which the need for invariance in respect to functional transformations has been overlooked is that the Confidence Interval, or range within which the parameter is not rejected on some test of significance, shall be as short as possible. This is an inappropriate requirement since the relative lengths of any overlapping intervals may be adjusted arbitrarily by a functional transformation.

A distinction without a difference has been introduced by certain writers who distinguish "Point estimation", meaning some process of arriving at an estimate without regard to its precision, from "Interval estimation" in which the precision of the estimate is to some extent taken into account. "Point estimation" in this sense has never been practised either by myself, or by my predecessor Karl Pearson, who did consider the problem of estimation in some of its aspects, or by his predecessor Gauss of nearly one hundred years earlier, who laid the foundations of the subject. The distinction seems only to be made in order to support a claim, which is not indeed historical, to the effect that the authors have made in this matter an original contribution. It shows great confidence in the ignorance of students to put such a claim forward.

The following is not a complete exposition of the Theory of Estimation, but an outline emphasizing the origin and relevance of the logical concepts used elsewhere in this book.

2. Criteria of estimation

The fundamental criterion of estimation is known as the Criterion of Consistency, and is essentially a means of stipulating that the process of estimation is directed to the particular parameter under discussion, and not to some other function of the adjustable parameter or parameters. Of the attempts I have made to express this idea, one at least has been quite unsatisfactory, and all perhaps deserve restatement.

If a number, finite or infinite, of observable classes have probabilities of occurrence

$$p_j\,, \qquad S(p) = 1\,,$$

which are known functions of the parameters; and if out of N observations, the numbers observed to fall in these are

$$a_j\,, \qquad S(a) = N\,,$$

then the average value of a linear function of the observed frequencies

$$A = S(c_j a_j) \tag{122}$$

when for each a is substituted its mean

$$\bar{a}_j = Np_j\,, \tag{123}$$

is given by

$$\overline{A} = NS(c_j p_j), \tag{124}$$

which is a known function of the probabilities, p, and therefore of the parameters.

The statistic A/N, calculable from the observations, will then be termed a Consistent estimate of the particular parametric function $S(c_j p_j)$. It will be noticed that it being linear in the frequencies is also an unbiased estimate of this same function.

Now, if, for example, the value of

$$S(c_j p_j) \,, \tag{125}$$

summed over all possible observational classes, were the parametric function

$$\log \theta \tag{126}$$

we might make the estimate

$$T = e^{A/N} \,, \tag{127}$$

and notice that when for all observed frequencies their expected values are substituted, the estimate T is such that it becomes identical with the estimand θ. This property is evidently invariant for transformations of the parameters, and does not imply the statement that T is an unbiased estimate of θ, for, in fact, it happens that $\log T$ is an unbiased estimate of $\log \theta$. It may be noted, for example, that the sufficient estimate T of Section 3, is not unbiased; and that this makes no difference to the fiducial distribution, of which the mean $\bar{\theta}$ is indeed equal to T.

In respect to bias it should be noted that no difficulty is usually experienced in adjusting an estimate so that the average of the adjusted value shall be equal to any particular parametric function; what has sometimes seemed to need emphasis is that the estimate is not necessarily improved by such an adjustment, which will introduce bias, not previously present, into the estimates of most functionally connected values. Before making any such adjustment, consideration should be given to its purpose.

The relations set out above are exact, and do not depend on the observed frequencies a being sufficiently large. It may well be that N of the frequencies are unity, and a very large, or infinite number, are zero; as, indeed will be the case when tolerably accurate measurements occur in the data, for these will be interpreted as single observations within very small ranges

$$x_1 \pm \tfrac{1}{2}\delta x_1 \,, \tag{128}$$

and any function

$$S\{c(x_i)\} \,, \tag{129}$$

where c is a continuous function of x, and the summation is taken over the observations and not over the classes, will be recognized as a linear function of the frequencies, and therefore as a suitable ingredient from which a consistent statistic can be built. The use of functions non-linear in the frequencies would in these cases introduce discontinuities whenever two measurements happen to coincide.

A Consistent Statistic may then be defined as:

> A function of the observed frequencies which takes the exact parametric value when for these frequencies their expectations are substituted.

This definition is applicable with exactitude to finite samples.

A much less satisfactory definition has often been used, namely that the probability that the error of estimation exceeds in absolute value any value ϵ, or, symbolically,

$$Pr\{|T-\theta|>\epsilon\} \,, \tag{130}$$

shall tend to zero as the size of sample is increased, for all positive values ϵ, however small.

With respect to a function of the observations, T, defined for all possible sizes of sample this definition has a certain meaning. However, any particular method of treating a finite sample of N_1 observations may be represented as belonging to a great variety of such general functions. In particular, if T' stand for any function whatsoever of N_1 observations, and T_N for any function fulfilling the asymptotic condition of consistency, then

$$\frac{1}{N}\{N_1 T' + (N-N_1)T_{N-N_1}\} \tag{131}$$

is itself a statistic defined for all values of N, and tending asymptotically to the limit θ, yet it is recognizable when $N=N_1$ as the arbitrary function T' calculated from the finite sample.

In fact, the asymptotic definition is satisfied by any statistic whatsoever applied to a finite sample, and is useless for the development of a theory of small samples.

3. The concept of efficiency

An asymptotic or large-sample definition is, however, appropriate as a first step in defining the concept of efficiency. Consider the statistic

$$T = \frac{1}{N} S(a_j c_j) , \tag{132}$$

in which the c_j are so far undetermined functions of θ, the parametric function of which T is to be an estimate. Then, T is consistent, when $\theta = \theta_0$, if

$$\theta_0 = S\{p_j c_j(\theta_0)\} , \tag{133}$$

and with the same coefficients c it will remain nearly consistent for small variations of θ if

$$1 = S\left\{\frac{\partial p_j}{\partial \theta_0} \cdot c_j(\theta_0)\right\} . \tag{134}$$

For large samples the variations to be expected may be made indefinitely small. Now the sampling variance of the linear function T is, exactly,

$$V(T) = \frac{1}{N} S\{p_j c_j^2 - \theta_0^2\}, \tag{135}$$

and this variance may be minimized for variations of the coefficients c, subject to the limitations that the estimate shall be locally Consistent, by minimizing

$$S(p_j c_j^2) - \lambda S\left(c_j \frac{\partial p_j}{\partial \theta_0}\right) - \mu S(p_j c_j) . \tag{136}$$

Varying any particular value c_i we find

$$2p_j c_j - \lambda \frac{\partial p_j}{\partial \theta_0} - \mu p_j = 0 , \tag{137}$$

from which μ may be determined by direct addition, giving

$$2\theta_0 = \mu , \tag{138}$$

and λ by multiplying by c and adding for all classes, giving a second equation,

$$2S(p_j c_j^2) = \lambda + \mu\theta_0 , \tag{139}$$

so that substituting for λ and μ in (137) for each particular coefficient

$$p_j c_j = \frac{\partial p_j}{\partial \theta_0}\{S(p_j c_j^2) - \theta_0^2\} + p_j \theta_0 , \tag{140}$$

or

$$p_j(c_j - \theta_0) = \frac{\partial p_j}{\partial \theta_0} S\{p_j(c_j - \theta_0)^2\} . \tag{141}$$

Hence it easily follows that

$$V(T) = \frac{1}{N} S\{p_j(c_j - \theta_0)^2\} = \frac{1}{NS\left\{\frac{1}{p_j}\left(\frac{\partial p_j}{\partial \theta_0}\right)^2\right\}}, \tag{142}$$

which we may write briefly as 1/I. Moreover,

$$c_j - \theta_0 = \frac{1}{p_j}\frac{\partial p_j}{\partial \theta_0} \div S\left\{\frac{1}{p_j}\left(\frac{\partial p_j}{\partial \theta_0}\right)^2\right\}. \tag{143}$$

For large samples, therefore, from populations having any particular parametric value θ, the linear function

$$T = \frac{1}{I} S\left(\frac{a_j}{p_j}\frac{\partial p_j}{\partial \theta_0}\right) + \theta_0\,, \tag{144}$$

is locally Consistent, and subject to this condition, has the least possible variance for samples of a given size. Since the p_j are known functions of θ, the general equation

$$S\left(\frac{a_j}{p_j}\frac{\partial p_j}{\partial \theta}\right) = 0 \tag{145}$$

or

$$S\left(\frac{a_j}{m_j}\frac{\partial m_j}{\partial \theta}\right) = 0\,, \tag{146}$$

if $m = pN$, being linear in the frequencies, though of any algebraic form in θ, will give, with large samples, such estimates of the highest precision for all values of θ.

In these limiting conditions the distribution of the estimate T tends to Normality, so that the specification of the variance as

$$\frac{1}{I} = \frac{1}{Ni}\,, \tag{147}$$

where, as above, i stands for

$$i = S\left\{\frac{1}{p_j}\left(\frac{\partial p_j}{\partial \theta}\right)^2\right\} = S\left\{p_j\left(\frac{\partial}{\partial \theta}\log p_j\right)^2\right\}, \tag{148}$$

is sufficient to specify fully the sampling distribution. The quantity I is the invariance of the estimate, and i

itself may be recognized as the amount of information to be expected for each observation made. The qualification "to be expected" is a reminder that the quantity i is itself a function of θ, of central importance also in the theory of small samples, and that with small samples the estimate of θ obtained from any sample will not be exactly correct.

The form of the Efficient equation of estimation,

$$S\left(\frac{a_j}{p_j}\frac{\partial p_j}{\partial \theta}\right) = 0 , \qquad (149)$$

shows that it could be derived by maximizing

$$S(a_j \log p_j) , \qquad (150)$$

or by maximizing that factor of the Mathematical Likelihood of θ which depends on θ, namely

$$\Pi(p_j^{a_j}) . \qquad (151)$$

The solution of this equation is Consistent, and invariant for transformations of the parameter. Naturally, it is not generally unbiased in the small sample sense.

The equations of Maximum Likelihood are indeed the only equations of estimation, linear in the observed frequencies, which are efficient with large samples. The solutions of these equations are not generally linear functions of the frequencies.

4. Likelihood and Information

The connections between the Likelihood Function and the Amount of Information are worth noting. The Likelihood Function is determined by a particular sample, or *corpus* of observations, and shows for such observations the relative frequency with which different parametric values would yield such a sample. When the logarithm of the likelihood is used, different independent samples, of the same or of different kinds, which throw light on the same parameter, may be combined merely by adding the log likelihoods for each value of the parameter.

For any particular value of the parameter, the probability of obtaining a given observational record may be represented by ϕ. Then using summation over all possible samples, it is clear that

$$S(\phi) = 1\,,\ S(\phi') = 0\,,\ S(\phi'') = 0\,, \qquad (152)$$

where differentiation with respect to a parameter is indicated by a prime.

Now

$$\frac{\partial^2}{\partial\theta^2}(\log\phi) = \frac{1}{\phi}\phi'' - \frac{\phi'^2}{\phi^2}\,, \qquad (153)$$

so that multiplying by ϕ and adding for all possible samples

$$E\left\{\frac{\partial^2}{\partial\theta^2}(\log\phi)\right\} = S(\phi'') - S\left(\frac{\phi'^2}{\phi}\right), \qquad (154)$$

or

$$0 - I\,,$$

since the last summation is merely the information expected for samples from populations having the particular value of the parameter chosen.

The average value of

$$-\frac{\partial^2}{\partial\theta^2}(\log\phi) = -\frac{\partial^2}{\partial\theta^2}L \qquad (155)$$

is thus equal to the amount of information expected. This relation shows very simply that the amount of information, like the likelihood function, is additive for independent bodies of data, even if of different sorts.

The value of the second differential coefficient of $(-L)$ with respect to θ is referred to as the amount of information realized at any value of θ. It is usually evaluated at that value for which L is maximized, since this has the highest likelihood of being the true value, but may be evaluated at any chosen value. At the maximum it measures the geometrical curvature. It is used in calculating the sampling variance of an estimate and is slightly more accurate, for this purpose, than the amount of information expected.

5. Grouping of samples

Ideally, we should like an estimate completely to replace the data from which it is drawn, so that no distinction need be made among different hypothetical lots of data which might yield the same estimate. This situation is sometimes, but not always, realizable.

If, for any given value of θ, the probabilities of observing the different types of samples which are to be grouped together are

$$\phi_1, \phi_2, \ldots \phi_s, \tag{156}$$

such that

$$\sum_1^s (\phi) = \Phi \tag{157}$$

Then

$$\sum_1^s \phi\left(\frac{\phi'}{\phi} - \frac{\Phi'}{\Phi}\right)^2 = \sum_1^s \left(\frac{\phi'^2}{\phi}\right) - \frac{\Phi'^2}{\Phi}, \tag{158}$$

and since the expression on the left cannot ever be negative, it follows that the contribution to the amount of information is never increased by such grouping, and that the condition that it shall not be diminished is that, for all configurations yielding indistinguishable estimates,

$$\phi'/\phi \tag{159}$$

shall be constant.

If this is so for all values of the parameter it follows that

$$\log \phi = L \tag{160}$$

shall be the same function of the parameter, apart from an additive constant. Samples to be grouped together must therefore have identical likelihood functions, if the grouping is not to be accompanied by loss of information. It is a characteristic of Sufficient statistics that the likelihood function of the statistic shall be the same as that of all samples from which such a statistic could have been calculated.

The actual loss incurred in other cases may be calculated by determining the frequency distribution of the statistic to be used in terms of its parameters, and calculating the amount of information supplied by a single observation from such a distribution. This supplies the criterion for efficiency in finite samples, and so completes the second stage of the theory. The information recovered will in general be less than the amount of information in the data from which the estimate was calculated, the differences having been lost through grouping the samples of different kinds which lead to the same estimate. With estimation by the Method of Maximum Likelihood, although the likelihood functions of samples leading to the same estimate may be different, yet the method of estimation has selected for grouping samples having likelihood functions so far alike as to have their maxima all at the same parametric value, and therefore with stationary ratios in the most important region. The method will necessarily lead to Sufficient Statistics when these exist.

In *The Theory of Estimation* (1925)[3] a good many examples have been given showing the loss of information for small samples of given size. When the likelihood function is everywhere differentiable these losses typically do not exceed the value of two or three observations; when the method of maximum likelihood is used in problems in which no sufficient estimate exists.

The concept of efficiency was introduced first in the foregoing discussion by a definition valid only in the limit for large samples. Such an approach is incomplete and must be taken only as a first step. In comparing estimates, all of which tend to be distributed Normally in the limit, a comparison of precision was immediately available; it is the evaluation of the maximum attainable that leads to the concept of the amount of information in an observation distributed in an error curve of any form, as is an estimate from a finite sample. It is the amount of information, and not the sampling variance, which completes the criterion of efficiency for finite samples, for when the distribution is not Normal the variance is an imperfect measure of the precision. The property, demonstrated above, that the amount of information may be diminished, or conserved, but cannot be increased by the processes of statistical reduction, guarantees the appropriateness of completing the definition by its means. Certain inequalities flow from the fact that a Normal distribution supplies the least possible information for an error curve of given variance, and has the least possible variance for a given amount of information.

6. Simultaneous estimation

In most practical situations there are a number of unknown parameters, not all of them necessarily of interest on their own account, but required in connection with the estimation of others. The likelihood is then a function of several variables, and is maximized by the solution of simultaneous equations, all linear in the frequencies, obtained by differentiating the log likelihood with respect to a complete functionally independent set of parameters. The solution is an efficient set of simultaneous estimates.

In the practical procedure of solution, trial values of the parameters are chosen, principally with a view to their being near to the final solution, so far as this can be foreseen, but partly also for their simplicity as a basis for calculation, and the expressions for

$$\begin{aligned} S_1 &= \partial L/\partial\theta_1\,, \\ S_2 &= \partial L/\partial\theta_2\,, \end{aligned} \qquad (161)$$

etc., are evaluated by substituting this particular set of parametric values. These are known as the efficient scores, and it is by their means that the trial value may be adjusted.

For a single parameter the adjustment is, as has been seen, effected by dividing the score by the amount of information, that is, if $\delta\theta$ is the adjustment required by a trial value θ', then

$$I\delta\theta = S \qquad (162)$$

will supply an adjusted value $\theta'+\delta\theta$ which is an Efficient estimate on the data available; for this reason it is seldom necessary, though always possible as a check, to repeat the calculation with an improved trial value.

With many parameters further polishing of the solution is more often wanted, because in some types of data, the first appraisal is less likely to be correct. The amount of information, which, with one parameter is a simple scalar, becomes a symmetrical matrix, or square table of coefficients, in the general case, the coefficients being

$$\left\{\begin{array}{lll} S\left\{\frac{1}{p}\left(\frac{\partial p}{\partial \theta_1}\right)^2\right\} & S\left\{\frac{1}{p}\frac{\partial p}{\partial \theta_1}\frac{\partial p}{\partial \theta_2}\right\} & \ldots \\ S\left\{\frac{1}{p}\frac{\partial p}{\partial \theta_1}\frac{\partial p}{\partial \theta_2}\right\} & S\left\{\frac{1}{p}\left(\frac{\partial p}{\partial \theta_2}\right)^2\right\} & \ldots \\ \text{etc.} & & \end{array}\right\} \tag{163}$$

or, written more compactly,

$$i_{rs} = S\left\{\frac{1}{p}\frac{\partial p}{\partial \theta_r}\frac{\partial p}{\partial \theta_s}\right\}, \tag{164}$$

where r, s, are suffices specifying particular parameters θ_r, θ_s.

The series of adjustments $\delta\theta$ are calculated from the linear equations of which the coefficients are given by the information matrix, and the right hand sides, by the Efficient Scores. In matrix form

$$I\delta\theta = S, \tag{165}$$

in which now $\delta\theta$ and S stand for series of values corresponding with the series of parameters.

An outstanding advantage of this form of analysis is that the precision of the simultaneous estimate is, for large-sample theory, given by a covariance matrix which is the reciprocal of the information matrix. So that if the matrix product IV is equal to the identity, then V supplies the variances and covariances of the components of the efficient estimate. The matrix V can then be calculated simply by replacing the scores by the several series

$$\begin{array}{llll} 1, & 0, & 0, & 0 \\ 0, & 1, & 0, & 0 \\ 0, & 0, & 1, & 0 \\ 0, & 0, & 0, & 1 \quad \text{etc.} \end{array} \tag{166}$$

and if this way is chosen the adjustments $\delta\theta$ are obtained by direct multiplication indicated by the matrix product

$$\delta\theta = VS .$$

It is to be noted that whereas the matrix I is exact for small samples, the identification of its reciprocal V with the covariance matrix of the simultaneous estimates is not exact; with small samples the sampling variation of the components will not generally be Normal, so that the covariance matrix cannot completely specify the distribution.

Properly speaking, therefore, the exact covariance matrix is not required; it could only be regarded as an approximation to V which does determine the simultaneous information matrix. To term V the covariance matrix is conventional and a useful reminder of an identity in large-sample theory.

With a single parameter, if θ were replaced by any known function $\phi(\theta)$, the amount of information, in each ingredient to be summed, is multiplied by

$$\left(\frac{\partial\theta}{\partial\phi}\right)^2$$

so that

$$I_\phi = \left(\frac{\partial\theta}{\partial\phi}\right)^2 I_\theta , \qquad (167)$$

where I_ϕ and I_θ stand for the amounts of information in any body of data in respect of ϕ and θ respectively. Similarly, with two or more parameters, when I has been replaced by a matrix which may still be written I_θ, if A is a matrix such that

$$a_{ij} = \frac{\partial\theta_i}{\partial\phi_j} , \qquad (168)$$

then I_ϕ is given by the matrix product

$$\begin{aligned} I_\phi &= A^* I_\theta A \,, \\ &= A I_\theta A^* \,, \end{aligned} \qquad (169)$$

where A^* stands for the transpose of A, and the two expressions are equivalent because I_θ and I_ϕ are symmetrical matrices.

This transformation system for the information matrices is exact with finite samples whereas the corresponding transformation of the covariance matrices could only be approximate.

Nevertheless, if

$$b_{ij} = \partial\phi_j / \partial\theta_i$$

the relation

$$v_\phi = B^* V_\theta B$$

is valid and exact, where V is the reciprocal of the information matrix.

7. Ancillary information

The study of the sampling errors, that is, of the precision, of statistical estimates, the core of the Theory of Errors as developed by Gauss, has led to the recognition among the multitude of Consistent estimates which can be invented, of a smaller class, such that, in the important class of cases in which the sampling distribution tends in large samples to the Normal form, then the limit of the product of the variance and the size of sample shall be as small as possible. The existence of such a limit has been demonstrated above, and its value has been expressed in terms of the relations between the unknown parameters and the frequencies of observations of all recognizable kinds. Such statistics of minimal limiting variances are termed Efficient, and thus satisfy a second rational criterion of what is required of a statistical estimate. It is easy to demonstrate[3] that any two estimates both efficient must have a correlation in random samples, and that their correlation coefficient must tend to a limit $+1$ as the size of the sample is increased. In fact, in the theory of "large samples" all efficient estimates are equivalent.

The theory of large samples can, however, never be more than a first step preliminary to the study of samples of finite size, although in fact a great many practical problems do not need for their effectual resolution, any further refinements. I do not think that this is a reason for not developing those concepts required for exact thought on small-sample problems. In such problems the different possible Efficient estimates must be distinguished. So far as the choice among them is concerned, a rational criterion is that we should prefer that estimate which conserves the most, or loses the least of the information supplied by the data. In practical terms, if from samples of 10 two or more different estimates can be calculated, we may compare their values by considering the precision of a large sample of such estimates each derived from a sample of only 10, and calculate for preference that estimate which would at this second stage give the highest precision. I do not know of a general proof, but no exception has been found to the rule that among Consistent Estimates, when properly defined, that which conserves the greatest amount of information is the estimate of Maximum Likelihood. The unique position of this method of estimation is also indicated by its being the only one in which the equations of estimation are linear in the frequencies. With small samples this obviates the irrationality of discontinuous changes in the estimates corresponding with minimal changes in the data.

A realistic consideration of the problem of estimation in small samples thus points unmistakably to the estimate of maximum likelihood as the uniquely appropriate single value for use in estimation, if any single value (i.e. with no ancillary values) is to be used. It indicates also that when using the maximum likelihood estimate, some loss of information may occur, and, although quantitatively this loss may be trifling, it cannot be unimportant logically, especially when an exhaustive treatment of the data is required in the calculation of probability statements.

The most important step which has been taken so far to complete the structure of the theory of estimation is the recognition of Ancillary statistics. The notion was first developed in a detailed study of the amount of information lost calculated exactly for a number of trivial but representative problems. It was shown not only that loss of information must vanish if all types of sample yielding the same estimate had identical Likelihood functions, and that for maximum likelihood estimates all must have functions stationary at the estimated value, but further that the loss of information suffered in the limit of large samples, was expressible directly in terms of the sampling variance of the second differential coefficient

$$\frac{\partial^2 L}{\partial \theta^2},$$

and indeed is equal in the limit to this variance multiplied by the variance of the estimate. When the likelihood function is repeatedly differentiable therefore, loss of information by simple estimation is due to the variance in the amount of information "realized". A simple remedy, appropriate for this asymptotic situation is merely to record, not only the estimate of maximum likelihood but the amount of information realized, or the "apparent precision", using this to supply a "weight" for the estimate more precise than that supplied by the size of the sample. Such a procedure, though merely asymptotic, and consisting of a refinement of the large-sample procedure, will guarantee that the loss of information in large samples shall tend to zero. The use of further differential coefficients of the Likelihood in the neighbourhood of the estimate can, by an extension of the same process, reduce the limit of N times the loss of information, or of N^r times its value to zero, as N tends to infinity. This is indeed an evasion of the problem of small samples properly speaking, but it does serve to show that it is the Likelihood function that must supply all the material for estimation, and that the ancillary statistics obtained by differentiating this function are inadequate only because they do not specify the function fully.

8. The location and scale of a frequency distribution of known form

If the probability of an observation falling into the range dx is given in the form

$$df = \exp\left\{\phi\left(\frac{x-\alpha}{\beta}\right)\right\}\cdot\frac{dx}{\beta}, \tag{170}$$

in which ϕ is a function of known form of the single real variable $(x-\alpha)/\beta$, which may be taken to be differentiable almost everywhere, and α and β are two unknown parameters specifying the location and the scale of the distribution, then the logarithmic likelihood function is

$$L = -N \log \beta + S\left\{\phi\left(\frac{x-\alpha}{\beta}\right)\right\}, \qquad (171)$$

where S stands for summation over the N values of a sample observed. It is not to be expected that the sum of the functions ϕ will be algebraically simple, and in consequence we shall not in general be able to express the likelihood in terms simply of the parameters and appropriate estimates of them. There will generally be no Sufficient pair of estimates.

The equations of maximum likelihood are easily seen to be

$$S\left\{\phi'\left(\frac{x-\alpha}{\beta}\right)\right\} = 0$$

$$S\left\{\frac{x-\alpha}{\beta}\,\phi'\left(\frac{x-\alpha}{\beta}\right)\right\} + N = 0, \qquad (172)$$

in which ϕ' stands for the first differential coefficient of the known function ϕ.

Let us now suppose that A and B are values of α, β satisfying the equations of estimation, and that ϕ is such that these values are real and unique, then the sample observed will have supplied a set of N values ot u such that

$$x_i = A + Bu_i\,, \qquad (173)$$

and the values of u satisfy the conditions

$$S\{\phi'(u)\} = 0 \qquad (174)$$

$$S\{u\phi'(u)\} + N = 0\,. \qquad (175)$$

The particular set of values u satisfying these equations, and derived from the sample observed, may be said to specify the complexion of the sample. It is easy to see that if for the values x observed we had instead the values

$$X = \lambda + \mu x \,, \tag{176}$$

then the complexion of the sample would be unchanged. To state the matter otherwise, the complexion depends only on the ratios of the $N-1$ successive differences among the N observations, when these are arranged in order of magnitude. It is evident, moreover, that the sampling variation of these ratios are severally and jointly independent of the parameters. The values u, specifying the complexion, thus define by their differences a set of $N-2$ functionally independent ancillary statistics, and the precision of the values A and B arrived at should be judged solely by reference to the variation of estimates among samples having the same complexion.

The precision of the estimates A and B may be specified by the measures of deviation

$$t_1 = \frac{A-\alpha}{\beta}\,, \qquad t_2 = \frac{B}{\beta}\,, \tag{177}$$

and since

$$\frac{x-\alpha}{\beta} = t_1 + ut_2 \,, \tag{178}$$

the simultaneous frequency distribution of t_1 and t_2 is

$$df \propto e^{S\{\phi(t_1+ut_2)\}}\, t_2^{\,N-2}\, dt_1\, dt_2 \,. \tag{179}$$

Moreover, the distribution of t_2 does not depend on α, being

$$df \propto \int_{-\infty}^{\infty} dt_1\, e^{S\{\phi(t_1+ut_2)\}}\, t_2^{\,N-2}\, dt_2 \,, \tag{180}$$

so that the fiducial distribution of β may be found independently of α, and thence that of α for given β, as in the case of the Normal distribution; in fact the simultaneous distribution of α and β in the light of a given sample is

$$df \propto \exp\left[S\left\{\phi\left(\frac{A-\alpha}{\beta}+u\frac{B}{\beta}\right)\right\}\right]\frac{B^{N-1}}{\beta^{N+1}}\,d\alpha\,d\beta\,, \quad (181)$$

for the actual set of u-values observed.

It can easily be verified that the distribution found in Chapter IV for the simultaneous frequency distribution of the mean (μ) and the standard deviation (σ) of a Normal distribution is a particular case of the solution given above, appropriate to the case

$$\phi(v) = c-\tfrac{1}{2}v^2\,; \quad (182)$$

the existence of Sufficient estimation, in that case, is replaced by the two estimates being in general rendered Exhaustive by taking account of $N-2$ independent Ancillary Statistics.

It is important to recognize the nature of the inversion that has been effected by the fiducial argument in this and in analogous cases. From the familiar form in which we have a frequency distribution of estimates such as A and B expressed in terms of the parameters, hypothetically supposed known, α and β, we have passed to a frequency distribution of α and β in a distribution specified by observable quantities including A and B. Since in reality A and B may be calculated from the observations, and are known in terms of them, while α and β are in reality unknown, the latter form of statement is the more realistic in representing the state of knowledge of a possible observer, while the former is a statement of what could be known in a hypothetical situation before any real observations had been made. By the fiducial argument we pass from the mathematical expression of a hypothetical to one of a realistic situation, in which the parameters are unknown, though exact probability statements can be made about them. Bartlett's criticism of the fiducial inference as "only to be regarded as a symbolic one" thus seems to be an example of mistaking the substance for the shadow. His statement that "there is no reason to suppose that from it we may infer the fiducial distribution of, say, $\mu + \sigma$" was presumably due to some analytic misapprehension. The problem involved no more than ordinary integration over the known bivariate frequency distribution, and its solution had been published before Bartlett wrote.

9. An example of the Nile problem

An example which illustrates well the connection between particular mathematical relationships on the one hand, and the existence of ancillary statistics, by means of which estimation can be made exhaustive, and exact probability statements inferred, as their consequences in mathematical logic, is as follows.

We suppose that pairs of observables (x, y) are distributed in a bivariate distribution

$$df = e^{-(\theta x + y/\theta)}\, dx\, dy\ , \qquad (183)$$

in which x and y take positive values only, then, if X, Y stand for the sums of the two coordinates over N pairs of observations, so that

$$S(x) = X\ , \qquad S(y) = Y\ , \qquad (184)$$

the Likelihood of any value θ in the light of the sample observed is

$$e^{-(\theta X+Y/\theta)} \tag{185}$$

so that its logarithm is determined by

$$L = -(\theta X+Y/\theta)\,. \tag{186}$$

The equation of maximum likelihood is then

$$X = Y/\theta^2\,, \tag{187}$$

leading to the estimate

$$T = \sqrt{Y/X}\,. \tag{188}$$

The amount of information expected from each pair of observations will be the mean of the square of

$$\begin{aligned} -\frac{\partial}{\partial\theta}&(\theta x+y/\theta) \\ &= \frac{y}{\theta^2}-x\,, \end{aligned} \tag{189}$$

but

$$E(x^2) = \frac{2}{\theta^2}\,,\quad E(xy) = 1\,,\quad E(y^2) = 2\theta^2\,, \tag{190}$$

hence

$$i_\theta = \frac{2}{\theta^2}\,, \tag{191}$$

or, considering the amount of information relative to log θ ,

$$i_{\log\theta} = 2\,. \tag{192}$$

The amount of information supplied in a sample of N values is therefore

$$\begin{aligned} I_\theta &= 2N/\theta^2 \\ I_{\log\theta} &= 2N\,. \end{aligned} \tag{193}$$

Since the likelihood cannot be expressed in terms only of θ and T, there will be no Sufficient estimate, and some information will be lost if the sample is replaced by the estimate T only. This loss of information may now be calculated from the exact sampling distribution of T.

10. The sampling distribution of the estimate

Since $2\theta x$ and $2y/\theta$ are distributed independently in exponential distributions equivalent to χ^2 for 2 degrees of freedom, it follows that $2\theta X$ and $2Y/\theta$ are similarly distributed for $2N$ degrees of freedom, or, stated otherwise that their simultaneous distribution is

$$df = \frac{X^{N-1}}{(N-1)!} \cdot \frac{Y^{N-1}}{(N-1)!} e^{-(\theta X + Y/\theta)} \, dX \, dY \, . \qquad (194)$$

Since we require the distribution of

$$T = \sqrt{Y/X} \qquad (195)$$

we may substitute

$$X = U/T \, , \qquad Y = UT \, , \qquad (196)$$

and obtain the simultaneous distribution of T and U, namely

$$df = 2e^{-U\left(\frac{T}{\theta} + \frac{\theta}{T}\right)} \cdot \frac{U^{2N-1} \, dU}{(N-1)! \, (N-1)!} \cdot \frac{dT}{T} \, . \qquad (197)$$

The distribution of T alone is obtained by integration with respect to U from 0 to ∞, giving

$$df = 2 \frac{(2N-1)!}{(N-1)! \, (N-1)!} \left(\frac{T}{\theta} + \frac{\theta}{T}\right)^{-2N} \frac{dT}{T} \, ; \qquad (198)$$

the logarithm of the factor involving θ is

$$-2N \log \left(\frac{T}{\theta} + \frac{\theta}{T}\right) , \qquad (199)$$

its differential coefficient with respect to θ is

$$\frac{2N}{\theta}\left(\frac{T}{\theta}-\frac{\theta}{T}\right)\div\left(\frac{T}{\theta}+\frac{\theta}{T}\right) \tag{200}$$

and the mean value of the square of this, giving the amount of information to be expected from a single observation, T, is to be evaluated.

Now from (198) it appears that

$$E\left(\frac{T}{\theta}+\frac{\theta}{T}\right)^{-2}=\frac{N^2}{2N(2N+1)}, \tag{201}$$

but

$$\left(\frac{T}{\theta}-\frac{\theta}{T}\right)^2=\left(\frac{T}{\theta}+\frac{\theta}{T}\right)^2-4, \tag{202}$$

hence

$$\left(\frac{T}{\theta}-\frac{\theta}{T}\right)^2\Big/\left(\frac{T}{\theta}+\frac{\theta}{T}\right)^2=1-4\left(\frac{T}{\theta}+\frac{\theta}{T}\right)^{-2}. \tag{203}$$

The amount of information is therefore

$$\frac{4N^2}{\theta^2}\left(1-\frac{4N}{2(2N+1)}\right)$$

$$=\frac{2N}{\theta^2}\cdot\frac{2N}{2N+1}, \tag{204}$$

being less than that supplied by the observations by one part in $(2N+1)$.

11. The use of an ancillary statistic to recover the information lost

The loss of information is less than half the value of a single pair of observations, and never exceeds one third of the total available. Nevertheless, its recovery does exemplify very well the mathematical processes required to complete the logical inference.

From the simultaneous distribution of U and T, we may find that of U only merely by integrating (197) with respect to T. The integral is in fact a standard form for the Bessel function K_0, and gives the distribution of U as

$$df = 4K_0(2U) \cdot \frac{U^{2N-1}\,dU}{(N-1)!\,(N-1)!}\,. \tag{205}$$

As this distribution is independent of θ, U is available as an ancillary statistic. The sampling distribution of T, taking U into account, is found by dividing the bivariate element by the corresponding marginal frequency of U, and is evidently

$$\frac{1}{2K_0(2U)}\, e^{-U\left(\frac{T}{\theta}+\frac{\theta}{T}\right)} \frac{dT}{T}\,. \tag{206}$$

From such an error distribution, having known U, the amount of information, calculated as usual, comes to

$$\frac{2N}{\theta^2} \cdot \frac{K_1(2U)}{K_0(2U)}\,, \tag{207}$$

which, it will be observed, depends upon the value of U actually available, but has an average value, when variations of U are taken into account, of

$$2N/\theta^2 \tag{208}$$

the total amount expected on the average from N observations; none is now lost.

The information is recovered and the inference completed by replacing the distribution of T for given size of sample N, by the distribution of T for given U, which indeed happens not to involve N at all. In fact, U has completely replaced N as a means of specifying the precision to be ascribed to the estimate. In both cases the estimate T is the same, the calculation of U enables us to see exactly how precise it is, not on the average, but for the particular value of U supplied by the sample.

In these circumstances it is possible to specify the precision by an exact statement of the probability of θ lying in any chosen range. Conveniently, if we write

$$\tau = \log T - \log \theta \,, \tag{209}$$

then τ has the distribution

$$\frac{1}{2K_0(2U)} e^{-2U \cosh \tau} d\tau \tag{210}$$

and the definite integral of this distribution between any chosen limits, τ_1 and τ_2 gives the probability that θ should lie between the corresponding limits

$$Te^{-\tau_1} \text{ and } Te^{-\tau_2} \,. \tag{211}$$

It will be noticed that the distribution of τ is symmetrical. The success of the process by which the missing information was recovered, and the statement of probability *a posteriori* rendered exact, evidently depends on the distribution of

$$U = \sqrt{XY} \tag{212}$$

being independent of θ. For a sample of one pair, it depends on the product xy having a distribution independent of θ, and therefore upon this being a solution of the Nile problem in the sense that the total frequency lying between any two rectangular hyperbolas

$$xy = c_1 \,, \qquad xy = c_2 \,, \tag{213}$$

shall be independent of θ, and depend only on the chosen values c_1 and c_2. Such curves therefore divide the total frequency in fixed proportions independently of the value of the unknown parameter, representing in that case the unknown height to which the Nile will rise.

With a different mathematical specification of the problem, different logical consequences might ensue. If we take a more general distribution

$$df = \theta\phi e^{-\theta x - \phi y}\, dx\, dy\,, \tag{214}$$

involving two parameters θ and ϕ, it may be seen that with any connection between θ and ϕ of the form

$$\phi = \theta^s \tag{215}$$

it will be possible to find an ancillary statistic and to derive probability statements about the parameter specifying θ and ϕ. However, the greater part of functional relationships which might subsist between two such positive quantities do not have this property, and apart from approximate statements appropriate to large samples, the totality of the information which the data supply is subsumed in the specification of the Likelihood function for all values of the unknown parameter.

12. Simultaneous distribution of the parameters of a bivariate Normal distribution

If from a Normal population with variances σ_1^2, σ_2^2 and correlation ρ, a sample yields the Sufficient estimates s_1, s_2 and r, then it was shown in 1915[2] that the sampling distribution of r was expressible in terms of ρ only, in the frequency element

$$\frac{1}{\pi(N-3)!}(1-\rho^2)^{\frac{1}{2}(N-1)}\,.\,(1-r^2)^{\frac{1}{2}(N-4)}\,.\,\frac{\partial^{N-2}}{\partial(\rho r)^{N-2}}\frac{\theta}{\sin\theta}\,.\,dr \tag{216}$$

where $\cos\theta = -\rho r$, and $0 \leqslant \theta \leqslant \pi$.

Since the distribution of r does not depend on the parameters other than ρ, we have a known function of r and ρ,

$$P(r,\rho)\,, \tag{217}$$

such that the distribution of r for given ρ is given by the frequency element

$$\frac{\partial P}{\partial r}\,dr\,, \tag{218}$$

and the frequency of ρ for given r is

$$-\frac{\partial}{\partial\rho}\{P(r,\rho)\}\,d\rho\,, \tag{219}$$

giving the marginal distribution of ρ in terms of r only. This was actually the first example of the derivation of a fiducial distribution (1930).[4]

For any given values of r, ρ the simultaneous distribution of s_1 and s_2 is

$$\frac{(N-1)^{N-1}}{(1-\rho^2)^{N-1}}\cdot\left(\frac{s_1\,s_2}{\sigma_1\,\sigma_2}\right)^{N-2}$$

$$\exp\left[-\frac{N-1}{2(1-\rho^2)}\left\{\frac{s_1^2}{\sigma_1^2}-2r\rho\,\frac{s_1\,s_2}{\sigma_1\,\sigma_2}+\frac{s_2^2}{\sigma_2^2}\right\}\right]\frac{ds_1\,ds_2}{\sigma_1\,\sigma_2} \tag{220}$$

divided by the function of $r\rho$ only

$$\frac{\partial^{N-2}}{\partial(-\cos\theta)^{N-2}}\frac{\theta}{\sin\theta}\,. \tag{221}$$

If we write

$$u=\sqrt{\frac{N-1}{1-\rho^2}}\cdot\frac{s_1}{\sigma_1},\quad v=\sqrt{\frac{N-1}{1-\rho^2}}\cdot\frac{s_2}{\sigma_2},\tag{222}$$

the distribution for given values of r, ρ becomes

$$(uv)^{N-2}\exp\{-\tfrac{1}{2}(u^2-2r\rho uv+v^2)\}\,du\,dv$$
$$\div\frac{\partial^{N-2}}{(\sin\theta\partial\theta)^{N-2}}\frac{\theta}{\sin\theta}.\tag{223}$$

Now for any chosen values ξ, η expression (223) will supply a function $P(\xi, \eta)$ such that

$$Pr\{u>\xi,\,v>\eta\}=P(\xi,\eta)\,,\tag{224}$$

or, dividing each value into

$$s\sqrt{(N-1)/(1-\rho^2)}$$

$$Pr\left\{\sigma_1<\sqrt{\frac{N-1}{1-\rho^2}}\cdot\frac{s_1}{\xi},\quad \sigma_2<\sqrt{\frac{N-1}{1-\rho^2}}\cdot\frac{s_2}{\eta}\right\}=P\tag{225}$$

giving the simultaneous fiducial distribution of σ_1 and σ_2 with the frequency element

$$\frac{\partial^2}{\partial\sigma_1\,\partial\sigma_2}P(\xi,\eta)\,d\sigma_1\,d\sigma_2$$
$$=\frac{N-1}{1-\rho^2}\cdot\frac{s_1\,s_2}{\sigma_1^2\,\sigma_2^2}\cdot(\xi\eta)^{N-2}\exp\{-\tfrac{1}{2}(\xi^2-2r\rho\xi\eta+\eta^2)\}\,d\sigma_1\,d\sigma_2$$
$$=(\xi\eta)^{N-1}\exp\{-\tfrac{1}{2}(\xi^2-2r\rho\xi\eta+\eta^2)\}\,\frac{d\sigma_1 d\sigma_2}{\sigma_1\sigma_2},\tag{226}$$

divided by

$$\frac{\partial^{N-2}}{\partial(-\cos\theta)^{N-2}}\cdot\frac{\theta}{\sin\theta},\tag{227}$$

and multiplied by the marginal frequency

$$-\frac{\partial}{\partial\rho}\{P(r,\rho)\}\,d\rho\,,\tag{228}$$

in which ξ and η are abbreviating symbols, standing for

$$\sqrt{\frac{N-1}{1-\rho^2}} \cdot \frac{s_1}{\sigma_1} \quad \text{and} \quad \sqrt{\frac{N-1}{1-\rho^2}} \cdot \frac{s_2}{\sigma_2} \cdot \tag{229}$$

The simultaneous fiducial distribution of these three parameters with two more for the population means, may then be found from the consideration that the set of statistics, s_1, s_2, r is distributed independently of the means, so that we need only a further factor representing the fiducial distribution of the means for known values of σ_1, σ_2 and ρ. Namely,

$$\frac{N}{2\pi\sqrt{1-\rho^2}} e^{\frac{-N}{2(1-\rho^2)}\left\{\frac{(x_1-\mu_1)^2}{\sigma_1^2} - 2\rho\frac{(x_1-\mu_1)(x_2-\mu_2)}{\sigma_1\sigma_2} + \frac{(x_2-\mu_2)^2}{\sigma_2^2}\right\}} \frac{d\mu_1\, d\mu_2}{\sigma_1\sigma_2} \cdot \tag{230}$$

It has been proposed that any set of functions having distributions independent of the parameters, such as

$$\left.\begin{aligned} t_1 &= \frac{s_1}{\sigma_1}\sqrt{N-1} \\ t_2 &= \frac{s_2}{\sigma_2}\sqrt{\frac{(N-1)(1-r^2)}{1-\rho^2}} \\ t_3 &= \sqrt{\frac{N-1}{1-\rho^2}}\left(r\frac{s_2}{\sigma_2} - \rho\frac{s_1}{\sigma_1}\right) \end{aligned}\right\} \tag{231}$$

can be used to transform the simultaneous frequency distribution of s_1, s_2, r in terms of σ_1, σ_2, ρ, into the simultaneous distribution of σ_1, σ_2, ρ in terms of s_1, s_2, r simply by multiplying by

$$\frac{\partial(s_1, s_2, r)}{\partial(t_1, t_2, t_3)} \times \frac{\partial(t_1, t_2, t_3)}{\partial(\sigma_1, \sigma_2, \rho)} \cdot \tag{232}$$

The example has been chosen to illustrate the process of building up the simultaneous distribution of the parameters rigorously by a step by step process, since the short cut suggested by the use of Jacobians has no claim to validity unless it can be proved to be equivalent to a genuine fiducial argument. The expressions (231) cannot indeed be made to supply such a proof. Perhaps the change of sign of $\partial t_2/\partial \rho$ at $\rho = 0$ should be a sufficient warning.

The correct way of using the facts stated in (231) has been more recently demonstrated by D. A. S. Fraser and D. A. Sprott who eliminate s_1, s_2 and σ_1, σ_2, obtaining the equation in three random variables

$$w = \frac{r}{\sqrt{1-r^2}}\chi_{N-2} - \frac{\rho}{\sqrt{1-\rho^2}}\chi_{N-1} \qquad (233)$$

in which w is a normal variable with mean zero and unit variance, while the two others are χ-variables with $(N-2)$ and $(N-1)$ degrees of freedom respectively. Assigning any value to ρ the equation gives the distribution of r as it was given in my paper of 1915, while assigning any value to r, it gives the fiducial distribution of ρ as first given in 1930, and as used in this chapter.

Explicit forms for the distribution of ρ have been derived by C. R. Rao from Fraser's formula. For example the form corresponding with (216) is

$$\frac{1}{\pi(N-3)!}(1-\rho^2)^{\frac{1}{2}(N-3)}(1-r^2)^{\frac{1}{2}(N-2)}$$

$$\frac{\partial^{N-3}}{\partial(\rho r)^{N-3}}\left\{\frac{\theta - \frac{1}{2}\sin 2\theta}{\sin^3\theta}\right\}d\rho \qquad (234)$$

REFERENCES

1. M. S. Bartlett (1937). Properties of sufficiency and statistical tests. *Proc. Roy. Soc.*, A, vol. 160, pp. 268-282.

2. R. A. Fisher (1915). Frequency distribution of the values of the correlation coefficient in samples from an indefinitely large population. *Biometrika*, vol. 10, pp. 507-521.

3. R. A. Fisher (1925). Theory of statistical estimation. *Proc. Camb. Phil. Soc.*, vol. 22, pp. 700-725.

4. R. A. Fisher (1930). Inverse probability. *Proc. Camb. Phil. Soc.*, vol. 26, pp. 528-535.

5. D. A. S. Fraser (1964). On the definition of fiducial probability. *Bull. Int. Statist. Inst.*, vol. 40, pp. 842-856.

INDEX